INSIDE THE HUMAN GENOME

INSIDE THE HUMAN GENOME

A Case for Non-Intelligent Design

John C. Avise

2010

OXFORD
UNIVERSITY PRESS

Oxford University Press, Inc., publishes works that further
Oxford University's objective of excellence
in research, scholarship, and education.

Oxford New York
Auckland Cape Town Dar es Salaam Hong Kong Karachi
Kuala Lumpur Madrid Melbourne Mexico City Nairobi
New Delhi Shanghai Taipei Toronto

With offices in
Argentina Austria Brazil Chile Czech Republic France Greece
Guatemala Hungary Italy Japan Poland Portugal Singapore
South Korea Switzerland Thailand Turkey Ukraine Vietnam

Copyright © 2010 by Oxford University Press, Inc.

Published by Oxford University Press, Inc.
198 Madison Avenue, New York, New York 10016

www.oup.com

Oxford is a registered trademark of Oxford University Press.

All rights reserved. No part of this publication may be reproduced,
stored in a retrieval system, or transmitted, in any form or by any means,
electronic, mechanical, photocopying, recording, or otherwise,
without the prior permission of Oxford University Press.

Library of Congress Cataloging-in-Publication Data
Avise, John C.
Inside the human genome : a case for non-intelligent design / John C. Avise.
p. ; cm.
Includes bibliographical references and index.
ISBN 978-0-19-539343-9
1. Molecular evolution. 2. Human genetics. 3. Genetic disorders.
4. Evolution—Religious aspects.
I. Title.
[DNLM: 1. Evolution, Molecular. 2. Genome, Human—genetics.
3. Genomics. 4. Religion and Science. 5. Selection (Genetics) QU 470 A958i 2010]
QH325.A895 2010
611'.01816—dc22 2009018050

1 3 5 7 9 8 6 4 2

Printed in the United States of America
on acid-free paper

QH
325
A895
2010

CONTENTS

PREFACE

Recent decades have been witness to the birth and growth of the Intelligent Design (ID) movement, the latest incarnation of a particular brand of religious creationism. Proponents of ID posit that complex biological outcomes, ranging from bacterial cells to human beings, were purposefully designed and directly crafted by a supreme intelligence (e.g., by a Creator God) rather than having arisen via nonsentient natural evolutionary forces. This has led to renewed calls that extra-natural explanations for life's origins and maintenance should replace—or at least be taught alongside—evolutionary biology in the science classrooms of public schools.

Life is indeed astonishingly complex, but explaining biological systems that are complex *and* function well is easy, at a cursory level. People of the relevant creationist persuasion need only invoke the attentive craftsmanship of a loving God, whereas the science-minded can always invoke the adaptive but unconscious agent of natural selection. Both a Creator God and natural selection are powerful shaping forces that can be expected to engineer complex biological features that operate smoothly and efficiently. The greater conceptual challenge—for scientists and ID proponents alike—is to explain complex biological traits that show profound structural or functional flaws.

Flawed biotic features nonetheless abound in species, including humans. Why—if either a Creator God or natural selection is in charge—do people have diabetes, heart attacks, bad backs, hemorrhoids, appendicitis attacks, difficult childbirths, impacted wisdom teeth, or any of a litany of other biological disorders? For science, the answers are not particularly mysterious—flaws abound because life has evolved stepwise under the influence of non-intelligent natural forces (including but not limited to natural selection, which itself has limited power). For some religions, however, the question can be more troublesome, and indeed it constitutes an age-old philosophical challenge known as the theodicy dilemma: Why do flaws and evil exist in a world run by a loving omnipotent God? Various religions have come up with provisional answers to the "problem of evil," such as the scenario in many branches of Christianity of how humanity fell from grace in the Garden of Eden. Other theologians have suggested that God for various reasons could not have created a world other than one run by natural laws, or, perhaps, that our world despite its many flaws is the best of all possible systems for the collective good.

Another possibility is that unpleasant and sometimes life-threatening biotic conditions (such as those listed above) are superficial phenomena that misrepresent an underlying perfection that God, or perhaps natural forces, engineered into molecular operations of the human genome (the full suite of genetic material within each of our cells). Some classical Greek philosophers, notably Plato, believed in the existence of two worlds: an eternal real world that is perfect and ideal, and an epiphenomenal world of imperfection that our senses perceive. Today, we understand that DNA (deoxyribonucleic acid) is the potentially immortal biotic component of our species, the key physical material passed from parents to progeny across the generations. Is it possible that the human genome is where biological perfection ultimately resides, and that any imperfect human conditions that emerge from that genome, or from interactions between DNA and the environment, are misleading—either about God's intent, or about the acumen of nature's design via unconscious evolutionary processes?

Here I hope to introduce readers with diverse educational and philosophical backgrounds to astounding and often unsuspected molecular features of the human genome and the relevance of these scientific findings to the question of underlying design. For millennia, theologians and biologists have pondered the human condition at the observable levels of anatomy and behavior, but only a half-century has passed since scientists began to study human genes more directly, and only a decade has transpired since scientists first sequenced a complete set of human DNA. Here we will delve again into the age-old mystery of the extent and source of human biological imperfections, but this time at the most fundamental molecular level. We will ask whether a God's benevolent intent might be divined by probing far beneath external or superficial appearances, that is, beneath the undeniable bodily frailties, behavioral flaws, and ethical shortfalls that are a conspicuous part of the human condition. Do molecular details inside the complex human genome finally provide theology's long-sought holy grail: direct and definitive evidence for attentive craftsmanship by a loving Creator God? Or do they point in a different direction?

In chapter 1—The Eternal Paradox—I discuss *the* recurring philosophical dilemma for theology: how to rationalize human suffering in a world governed by a loving omnipotent deity. I review the history of the nuanced relationship between religion and science over matters of perfection and imperfection in biology, and I introduce the numerous explanations for biological flaws that have emerged from theology and science. I explain how the Darwinian revolution that began in the mid-19th century eventually transformed the life sciences analogously to how the Copernican revolution transformed physics and astronomy three centuries earlier. I also review the history and perspectives of Intelligent Design, and highlight several misconceptions that the ID movement often promotes about the nature of evolutionary processes.

The next three chapters delve into the many genetic malfunctions and apparent flaws of molecular design that characterize the human genome. Chapter 2—Fallible Design—addresses the history and current state of knowledge about human "inborn errors of metabolism," that is, how mutations in protein-coding genes kill countless individuals (especially embryos and fetuses) and promote legions of genetic disabilities in our species. Chapter 3—Baroque Design— addresses various gratuitous genomic complexities that routinely compromise human health. These range from problems associated with split genes and complications from the Byzantine mechanisms of gene regulation and nucleic acid surveillance, to the peculiarities of genomic imprinting and the bizarre structure and function of mitochondrial DNA. Chapter 4—Wasteful Design—treats repetitive elements in the human genome, ranging from duplicons and pseudogenes (dead genes) to several classes of ubiquitous mobile elements that look and act much like intracellular viruses. For each category of genomic feature discussed in chapters 2–4, I contrast various theological excuses for the flaws with explanations that come from evolutionary biology.

Chapter 5—Intelligent or Non-Intelligent Design?—recapitulates evidence for rampant imperfection in the molecular architecture and operations of human genomes. Inherent structural flaws as well as mutational glitches combine to produce genomes that are complex but often dysfunctional and that invariably fall far short of designer perfection. These genetic findings extend the age-old theodicy challenge into the previously unexplored molecular realm. From a scientific vantage, however, such genomic flaws are hardly surprising, because natural selection is nonsentient and far from all powerful. Contrary to ID predictions, the human genome is not irreducibly complex, and the inevitable imperfections it displays are consistent with evolutionary expectations.

The ID movement has framed issues about biological causation in ways that needlessly pit science against religion. The molecular blemishes described in this book, interpreted in the light of evolution, motivate a new perspective on how mainstream religions could rightly view the evolutionary sciences—not as damnable antagonists in explaining the flawed human condition but rather as serviceable if not philosophically uplifting allies. Indeed, with respect to some of the most intimate molecular details of our flawed physical existence, evolution could be interpreted as a philosophical savior for theology by helping to emancipate the latter from the age-old shackles of theodicy.

I would like to extend my sincere thanks to Joan Avise, Francisco Ayala, Felipe Barreto, Rosemary Byrne, Loren Rieseberg, Michael Ruse, Andrey Tatarenkov, Vimoksalehi Lukoschek, my editor Peter Prescott, and several anonymous reviewers for constructive comments on earlier drafts of the manuscript. Of course, any errors of fact or interpretation remain my responsibility.

INSIDE
THE
HUMAN GENOME

CHAPTER 1

The Eternal Paradox

The year 2001 was extraordinary. At the gateway of a new century and millennium, it began with a renewed sense of hope for the near-term and longer term future. Here was a fresh chance for humanity to distance itself from the endless parade of wars and atrocities that had characterized much of the 20th century and before. Here was an unrivaled opportunity to reap the benefits of powerful biological sciences and new genetic technologies in such areas as medicine, agriculture, energy, conservation, and ecological sustainability. Here perhaps was the dawn of a new era when unprecedented scientific discoveries might be interwoven with cherished religious or philosophical traditions to yield richer fabrics for comprehending humankind's place in any broader scheme. Sadly, such optimism was quickly overshadowed by other world developments.

The year had started superbly with the announcement in mid-February of a momentous scientific achievement. Two independent teams of geneticists that had been working in spirited competition for several years simultaneously published the first draft sequences of the human genome[1] (the full set of DNA within a human cell). This landmark accomplishment was a gigantic first step toward hoped-for breakthroughs not only in medicine but also, perhaps, in humanity's long quest for self-understanding. By reading the set of genetic files, containing the complete molecular blueprints of our species on all 23 pairs of chromosomes, these research teams in effect had launched a scientific exploration into the subcellular world of inner space that promised to be at least as intellectually engaging as the science-fiction exploration of outer space that Stanley Kubrick and Arthur Clarke had portrayed in their science fiction film classic, *2001: A Space Odyssey*.

Then came the horrific acts of September 11, 2001. Once again, the human capacity for inhumanity, associated so often with religious fanaticisms of one sort or another, had reared its ugly head. George W. Bush, himself an evangelical Christian, appropriately denounced these terrorist atrocities, and then identified an Axis of Evil where such acts purportedly were state sponsored. The United States reacted to the despicable events of 9/11 by launching wars in Iraq and Afghanistan, where we have been painfully reminded that nastiness and brutality still abound in the world, that civilization may be only a thin veneer over barbarism, and that fundamentalist attitudes and traditional sociopolitical habits are hard to break. The events of 2001—from the Human Genome Project to 9/11—powerfully illustrate the deep gulf that exists between the most noble and the most banal of human activities.

THEODICY

The question arises after every natural tragedy. It issues from the lips of survivors often as whispers, sometimes as agonized screams. It is thought or spoken by the young and the old, by the rich and the poor, by the erudite and the uneducated. Throughout history, it has been our emotional response to grievous misfortune. As we helplessly witness the aftermath of a flood's brutality, or that of a hurricane, tornado, tsunami, earthquake, famine, or pestilence, we cry out for answers. Desperate for any modicum of comprehension and relief, we plead for reassurance and solace from a higher source. And we feel compelled to ask, *why*? Why do You,

Almighty God, countenance the horrific events that produce such misery?

I write these words in the autumn of 2005. Within the past few months, the world has been witness to a tsunami of epic proportions that killed more than 200,000 people in and around Indonesia. It has seen devastating hurricanes and floods that left hundreds dead and thousands homeless in the southern United States and Central America, and it has beheld a catastrophic earthquake that claimed more than 80,000 lives in northern Pakistan. Yet, such mass calamities are only the most conspicuous sources of human misery. Personal tragedies occur continually and take a far greater collective toll. To a senior citizen who has just lost her lifelong companion to Alzheimer's disease, to a child newly orphaned by an automobile accident, to the parent of a son or daughter lost in military combat, to the bereaved mother of a stillborn infant, or to anyone in desperate grief or agony, an overwhelming question often arises: How can a merciful God be so cruel? Jesus himself, before his crucifixion, is said to have cried out, "my God, my God, why hast Thou forsaken me?"[2]

This is *the* persistent dilemma of theology. Why does suffering exist in a world governed by a loving deity? Put more simply, why do bad things happen to good (as well as bad) people?[3] Some of the proximate sources of human misery are geophysical or meteorological events that apologists might suggest lie outside a Creator God's chosen jurisdiction. Earthquakes and floods exemplify what sometimes is termed natural evil, in distinction to moral evil which seems to be instigated by humans. As was made abundantly clear by the crucifiers of Christ, and by the 9/11 terrorists, many of humanity's trials and tribulations are self-inflicted moral evils, by humans on humans. People routinely disobey the Golden Rule at scales ranging from relatively innocuous to egregious. They steal, lie, and commit murders and rapes; and nations wage wars and commit genocides. Why does God allow children made in His image to behave so abominably?

Even if this behavioral quandary could somehow be answered (perhaps as the inevitable price of the free will that God has granted his flock), the broader issue would remain as to why vast human misery stems from debilitating frailties of our own physical makeup (including our genes) over which the afflicted presumably have no control. Why for example must countless people endure sometimes lifelong pain and suffering from inherited physical disabilities? And why must the persecutions extend even to those who surely must be innocent in God's eyes, such as the untold numbers of human embryos and fetuses who spontaneously abort in utero? Indeed, in a world run by a loving Creator God, why are senescence and death the unavoidable fates of anyone fortunate enough to have survived all of life's earlier challenges? Are free will and the opportunity for divine redemption the best explanation for such traumas and turmoils of mortal existence?

Translated literally, the term theodicy (from the Greek roots theós for God and dik for justice) means "God's justice." The German philosopher and mathematician Gottfried Leibniz coined the word in 1710 in a work entitled *Theodicic Essays on the Benevolence of God, the Free Will of Man, and the Origin of Evil.* Theodicy describes philosophical attempts to vindicate God's holiness and justice in establishing a world that is rife with physical and moral woes. Theologians and others have tried to rationalize this monumental "problem of evil" in modes ranging from supernatural revelation and faith, to attempted interpretations of scientific evidence by objective logic.

GOD'S WILL BE DONE

Ever since the dawn of human self-consciousness, people have sought to comprehend the sources of humankind's many curses.[4] Rationalizations from theologians and other philosophers are legion. Perhaps evil in the world stems from free will that a righteous God purposefully ordained so that people can choose or not to follow God, or perhaps so that human souls can be judged worthy of reward in heaven or eternal damnation in hell according to specific actions that each individual chooses in life. Under this common Christian view, free will itself is interpreted as a necessary and sufficient manifestation of intelligent design by a Creator God. Thus, some Christian theologians have concluded that God could not have created a world other than that run by natural laws because to do so would have been incompatible with humans being substantially free moral beings.

But what about the evil visited upon people in situations where free will is irrelevant? The victim of a tornado or earthquake presumably had no control over such natural catastrophes, nor did an aborted embryo have control over the inherited disabilities that may have caused its premature death. Perhaps God oversees these happenings directly, or perhaps He merely set the physical and biological stages on which these dramas play out according to more natural scripts. Perhaps the human condition, despite its many flaws, is the best of achievable worlds, with any alternatives being even less favorable with respect to the collective good. The broader point is that philosophers—including Christian and other theologians—have advanced many suggestions but have reached no universal consensus about the problem of evil.

The number of theistic explanations for evil and human suffering probably surpasses the number of different Creator Gods (tens of thousands) that various cultures have worshipped over the ages because each such God can be attributed multiple motives and courses of action. Many of the popular rationales for suffering in the world are categorically similar, even if different in detail. Leibniz suggested that God cannot do the impossible and that what He created is the best possible world, with evil as an unavoidable part of the package deal (as illustrated, for example, by the pain of burning balanced by the good of avoiding fire). In other words, any other realized outcome would have entailed even more evil.

Many other theological rationales have been proposed. Perhaps God allows natural processes to operate unfettered to prove that happenings outside His immediate direction can indeed be abhorrent. Perhaps God dictates human suffering as trials or challenges that encourage better ethical qualities (such as patience, empathy, obedience, or piety). Perhaps God dictates death as the necessary prelude to an eternal afterlife in Heaven or Hell. Perhaps persons afflicted with pain and suffering in this life are paying a toll for humankind's sins in past existences. Perhaps each of us, regardless of personal character, must pay a painful price for the collective sin currently on display in our species. Perhaps God's concepts of good and evil differ basically from our own, and much of what we perceive to be evil is actually good.

Maybe Satan counters God's good intentions, and succeeds on occasion. Maybe an all-powerful and loving Creator God died. Maybe the Creator God became senile, distracted, disgusted, overwhelmed, lackadaisical, clumsy, mischievous, or diabolical, or maybe He simply became disinterested in human affairs. Maybe God is a masochist and is satisfied or entertained by human suffering. Perhaps multiple Gods exist and they designed life sloppily, by committee. Maybe pain and evil are illusory, mere figments of our imagination. Such musings are endless, and no such philosophical conjecture can be accepted or rejected except as an article of faith. These kinds of theological hypotheses appear not to be objectively testable on an evidentiary basis. Maybe a Creator God does not want us to test such hypotheses because He is vain, embarrassed, wishes to remain mysterious, or has any other excuse that you might imagine.

Some theologians suggest that God's reasons for permitting evil are valid but beyond human comprehension (in which case there is really no point in considering the matter further). They suggest that we can have no idea of what God had in mind when he created life. This argument, which often springs up in the defense of Creation Science and Intelligent Design, presents a philosophical conundrum, as explained by Elliott Sober, a philosopher of biology:

> the problem with the design hypothesis is that we have no independent knowledge of the goals and abilities that the designer of organisms would have if such a being existed... [and earlier]...the problem with the hypothesis of intelligent design is not that it

makes inaccurate predictions but that it doesn't predict much of anything.[5]

Another possibility is that God is nothing more nor less than natural forces and natural processes themselves. The famous evolutionary biologist Theodosius Dobzhansky once wrote, "I am a creationist *and* an evolutionist. Evolution is God's, or Nature's method of creation";[6] and the respected architect Frank Lloyd Wright similarly stated, "I believe in God, only I spell it Nature." This is the sort of deity (rather than a Creator God directly involved in daily human affairs) that Albert Einstein and many other physicists have sought to understand in their explorations of light, energy, matter, and the structure of the universe.[7] It is also the kind of God that many scientists and philosophers (especially since Darwin) have tried to comprehend within the sphere of biology. When exploring the nature of life, including the human condition, these scholars generally try to avoid both subjective and metaphysical explanations, and focus instead on more assessable propositions based on best attempts at impartial empiricism and dispassionate logic.

All scientific hypotheses and theories remain provisional (by the definition of science), forever subject to critical evaluation and possible reinterpretation with new evidence. All basic truths within a fundamentalist religion remain eternally valid (by the fundamentalist definition of religious truth), forever immune from possible falsification by empirical experience or objective reason. Thus, religious fundamentalists tend to perceive science's primary strength (evidence-based adjustability) as a profound weakness, and secularists tend to perceive religion's primary strength (faith-based certainty) as a fundamental flaw. Because religion tends to be grounded in faith that can be inimical to scientific inquiry, whereas science is grounded in skepticism that can be anathema to religious faith, it is little wonder that these two epistemological approaches often seem at odds with regard to matters of ultimate concern.[8] But a far more nuanced relationship often has existed between science and religion, and nowhere has this been more evident than in the history of mankind's attempts to interpret the many complexities and exquisite beauties (the seeming antithesis of biological imperfection) in nature's design.

REVOLUTIONS OF PLANETS, AND OF HUMAN THOUGHT

The Middle Ages in Europe (from about 500 A.D. to the Age of Enlightenment in the 1700s) were often interpreted by historians of the 18th and 19th centuries as a time when church doctrines and religious faiths, antithetical to science, reigned supreme. In particular, the years 500–1000 A.D. were later assigned the pejorative title of Dark Ages[9] because that period left few written records and presumably spawned relatively few cultural achievements. However, many modern historians now appreciate that at least some of the appearance of darkness and its contrast with the later Enlightenment were in the eyes and prejudices of beholders, and that the realities (both then and now) were far more subtle. For example, in many parts of the world even today, religious beliefs that sometimes contradict modern science still hold primary sway over people's concepts and actions. Furthermore, religion in the Middle Ages did not always conflict with the science of that era. Indeed, a strong argument can be made that Christianity in Europe (like some other religions elsewhere)[10] was actually the primary *motivation* for a powerful mode of empirical inquiry known as natural theology (see below).

Nevertheless, two major transformations of thought, one in the Middle Ages and one in the post-Enlightenment era, were to become huge sources of conflict between scientific outlooks and some long-established brands of religious fundamentalism.

The Copernican Revolution

In 1514, Poland-born Nicolas Copernicus circulated first outlines of his heliocentric theory that challenged prevailing Earth-centered (and ergo human-centered) interpretations of the universe. Using mathematical arguments based on detailed observations of planetary movements, Copernicus concluded that the Earth and several other celestial bodies revolve around the sun and thereby constitute a solar system. Furthermore, he suggested, the heavens must be rife with similar physical arrangements. The ramifications were profound: no longer could the Earth and its human inhabitants be viewed quite so comfortably as the central focus of all Creation.

During the Middle Ages, especially in Europe, such attempts to interpret the cosmos or other physical phenomena via empiricism and mathematical deduction were rare and generally dismissed or even damned[11] by the Church when they yielded outcomes apparently contrary to the Scriptures. Thus, the most salient feature of the Copernican revolution was not the proof of heliocentrism per se (important though that was) but rather the powerful illustration of a then-radical mode of intellectual inquiry that was unshackled from sacred texts or religious revelations.[12] What would now be called the physical sciences (including physics and astronomy) had begun to wrest from theologians and religious authorities at least some degree of explanatory power over mechanistic operations of the universe.

Natural Theology

Three more centuries would pass, however, before an analogous Darwinian revolution would likewise transform the biological sciences. In the interim (as well as in many prior ages), a common sentiment was that religious inquiry and biological inquiry were strong intellectual allies rather than adversaries in missions to explain and thereby glorify God and His Creation. Scientists and religious leaders often shared a conviction that the careful study of nature would only confirm God's unique invention and oversight of life. In those more comfortable theological times, most leading scientists (i.e., "natural philosophers") were avowed deists, and many leading clerics were also notable science-scholars. All were busily engaged in confirming God through rational inquiry, which frequently was seen as a helpful if not necessary complement to traditional knowledge of God from gospel truths and religious revelations. When objections to science were raised in theological circles (and they often were), they usually came from a fearful notion that it might be heretical or even dangerous (given a wrathful God) to try to prove empirically that which required no proof: God's magnificence as detailed in the Scriptures.

In 1802, Reverend William Paley published a treatise, *Natural Theology*,[13] that formally explained what many field naturalists and other biologists had for centuries consciously and explicitly sought to accomplish in their studies of nature: final proof and glorification of God's majesty through empirical analyses of His works. These natural theologians typically started with two basic premises: that life's beauty and complexity were prima facie evidence of God's creative power, and that by carefully analyzing living nature they inevitably would exalt God, and perhaps also come better to comprehend His purposes.

Paley's most famous passage illustrating the "argument from design" began by supposing that while traversing a heath, hikers had discovered a fine watch lying upon the ground, and that they felt philosophically inclined to contemplate its source:

> When we come to inspect the watch, we perceive, (what we could not discover in the stone), that its several parts are framed and put together for a purpose, e.g. that they are so formed and adjusted as to produce motion, and that motion so regulated as to point out the hour of the day; that, if the several parts had been differently shaped from what they are, of a different size from what they are, or placed after any other manner, or in any other order... none which would have answered the use, that is now served by it... This mechanism being observed... the inference, we think, is inevitable; that the watch must have had a maker; that there must have existed, at some time and at some place or other, an artificer or artificers who formed it for the purposes, which we find it actually to answer; who comprehended its construction, and designed its use.

The analogy, of course, is between the watch's interacting gears, springs, and sprockets (obviously constructed by humans) and the even more complex and interdependent tissues and organs of a living creature (which therefore must have been designed purposefully by an intelligent artificer, God). Another analogy is to liken the watch to a natural ecosystem in which interacting species play their respective roles in a complex community that could have been ordained only by an intelligent Creator. All sophisticated biological structures or functions were interpreted as ineluctable proof for God's immediate directive hand.

Paley's treatise was the most thoughtful and influential of several summaries of natural theology published shortly prior to the Darwinian revolution. Others included the Bridgewater Treatises, a series of eight major works published between 1833 and 1840.[14] However, many scholars in earlier times also had espoused the idea that biological complexity evidences God's design. Four centuries before Christ, Plato mused extensively in his dialogues (which often purported to involve conversations between Socrates and his followers) about the complexity of the world and about how a conscious mind (a Designer God) must be behind it all. And in 1691, the Christian scholar and scientist John Ray extended such thoughts in *The Wisdom of God Manifested in the Works of Creation.* Thus, although many of the arguments in *Natural Theology* were eloquent and the author himself was a highly knowledgeable biologist for his era, Paley was certainly not the first (nor the last) of the natural theologians.

Even Paley's analogy involving a finely crafted watch had been employed earlier, perhaps first by the Roman philosopher Marcus Cicero a century before Christ. In *De Naturae Deorum* (*On the Nature of the Gods*), one of Cicero's characters asks,

> Suppose a traveler to carry into Scythia or Britain the orrery [a clocklike model of the solar system] recently constructed by our friend Posidonius, which at each revolution reproduces the same motions of the sun, the moon and the five planets...would any single native person doubt that the orrery was the work of a rational being?[15]

In substance, Paley's argument from design was also just a restatement of one of the so-called Five Ways[16] that St. Thomas Aquinas (an influential 13th-century Dominican scholar) purportedly had proved the existence of a Creator God. Indeed, natural theology was simply common sense, even to the uneducated. As phrased by John Ray in *The Wisdom of God,*

> You may hear illiterate persons of the lowest Rank of the Commonalty affirming, that they need no Proof of the being of God, for that every Pile of Grass, or Ear of Corn, sufficiently proves that.... To tell them that it made it self, or sprung up by chance, would be as ridiculous as to tell the greatest Philosopher so.

Such was the conceptual climate in which most biological observations in Europe were interpreted during the Middle Ages (and often before). However, even more novel and exciting opportunities to explore God's creation were soon to present themselves. At about the time of Copernicus, European adventurers had discovered the Americas and begun to explore these and other distant lands, which often teemed with many types of plants and animals that Europeans never before had seen. Many naturalists in the ensuing three and a half centuries rose to the challenge of describing these unanticipated finds and explicating them in the context of God's plan. For example, in the diaries of his travels across southeastern North America in the late 1700s, William Bartram provided detailed descriptions of a rich native flora and fauna, which he interpreted as constituting "a glorious apartment of the boundless palace of the sovereign Creator,...furnished with an infinite variety of animated scenes, inexpressibly beautiful and pleasing."[17]

A century later, when similarly documenting his wilderness journeys across western North America, the naturalist John Muir expressed his sentiments about nature's bounty thus:

> the hills and groves were God's first temples;...every crystal, every flower a window opening into heaven, a mirror reflecting the Creator;...all the wilderness seems to be full of tricks and plans to drive and draw us up into God's Light;...and...I thank God for this glimpse of it.[18]

In the early 1700s, as reports and collections of newly discovered organisms from around the world streamed back to Europe, the Swedish biologist Carl Linnaeus set before himself the worthy task of naming and cataloguing all species from "God's garden" into a hierarchical scheme of classification (much like a librarian might organize books by topics and subtopics). Linnaeus was to become the universally acknowledged father of biological taxonomy, and his general nomenclatural framework for systematics is still followed today. He was also an avid naturalist as well as a biblical creationist who believed in the near-fixity of species. He often admitted finding himself "completely stunned by the resourcefulness of the Creator....I saw the infinite, all-knowing and all-powerful God from behind....I followed His footsteps over nature's fields and saw everywhere an eternal wisdom and power, an inscrutable perfection." Linnaeus inscribed the frontispiece to his 1753 magnum opus—*Species Plantarum*—with a passage that he embellished from the Old Testament (Psalms 104:24): "O Jehovah, how ample are Thy works! How wisely Thou hast fashioned them! How full the earth is of Thy possessions."

Meanwhile, on a very different empirical front, the invention of magnifying lenses and primitive microscopes in the early 1600s permitted biologists to gain first glimpses into the heretofore unsuspected world of biotic minutiae. In 1665, after examining thin slices of cork (from oak bark) at 30X magnification, the Englishman Robert Hooke described and named "cells." Just a few years later, the Dutch microscopist Antony van Leeuwenhoek (who had been born into a Dutch Reformed tradition that included a firm belief in biblical creation) published a series of scientific reports on forms of life examined at magnifications as high as 300X. His observations included the first identification of sperm cells and red blood cells, the first detailed descriptions of tiny body parts such as muscle striations, optic lenses, and fine structures of insect mouthparts, and the discovery of various "animalcules" including microscopic algae, rotifers, protozoa, and bacteria. In his writings, Leeuwenhoek clearly took great delight in interpreting such findings as confirmation of God's attention to detail:

> From all these observations, we discern most plainly the incomprehensible perfection, the exact order, and the inscrutable providential care with which the most wise Creator and Lord of the Universe had formed the bodies of these animalcules, which are so minute as to escape our sight, to the end that different species of them may be preserved in existence. And this most wonderful disposition of nature with regard to these animalcules...which at the same time strikes us with astonishment, must surely convince all of the absurdity of those old opinions, that living creatures can be produced from corruption of putrefaction.[19]

Thus, as European naturalists spanned out across the globe and peered into microscopes during the 15th through 19th centuries of biological discovery, they encountered vast new wonders of nature, many more evidences of the bounty and grace of God's Creation. More often than not, Christianity (even fundamentalist Christianity) had inspired these natural theologians. Indeed, it is doubtful that some biologists would have undertaken such arduous empirical studies with due zeal and diligence had it not been for the added spur of religious motivation.

Ironically, however, biological discoveries by the natural theologians also had begun to open a Pandora's box of troubling questions for particular sects of the Christian Church. Why, if God had created life in one place just a few thousand years earlier, did organisms differing in kind and in degree now occupy different parts of the world? How could individuals representing all species have fit onto Noah's Ark? Why were microbes part of Creator God's plan, especially those that soon (starting in the late 1800s) were discovered to cause dreadful human diseases such as smallpox and bubonic plague? By looking outward to biotas on other continents, and inward to microscopic features of life, European biologists had begun to encounter difficulties not only with literal readings of Genesis but also with what seemed to be troubling contradictions in the guiding philosophy of natural theology, which had presumed that God's perfection pervades all of nature's design.

Another emerging difficulty for the Church, involving the temporal dimension of God's Creation, was presented by the strange castes and imprints often displayed in sedimentary rocks. These inanimate structures of evident design certainly looked lifelike yet did not always closely resemble living species. Philosophers at least since Greek and Roman times had known of fossils and sometimes concluded that they must represent once-living organisms that had been turned to stone. But nearly two millennia later, how to interpret these natural curiosities was still a matter of serious debate. For example, in 1671 a paper was presented at England's prestigious Royal Society in which a leading Oxford scholar (Martin Lister) proposed that fossils were not evidences of past life but instead were physical artifacts of a "plastick virtue" inherent in rocks. This notion eventually waned as scientific evidence about the actual nature of fossils gradually accumulated.

Fossils themselves had been of little concern to biblical scholars so long as they were interpreted merely as interesting trinkets of nature, but as their biological sources and geological ages became better known, they too posed growing theological dilemmas for biblical literalists. How could ancient fossils be reconciled with a fixed and recent Creation? And, even if the emerging sciences of biology and geology were completely wrong about the antiquity of the planet and its fossils, why had God taken the trouble to create creatures that He then allowed to go extinct?

In summary, Christianity's natural theology had been a primary intellectual sponsor for the rise and spread of the biological sciences in Europe during and immediately following the Middle Ages.[20] Ironically, however, the science spawned by natural theology had also begun to confront its benefactor with empirical findings that often eroded the theistic beliefs (in an omnipotent God who directs all of creation) upon which the natural sciences themselves had been founded.

The Darwinian Revolution

When Charles Darwin stepped aboard the H.M.S. *Beagle* in 1831, initially as a traveling companion to Captain Fitzroy and eventually to become the ship's naturalist, he too was a natural theologian enthusiastic about describing the Creator God's works. Having failed to complete his earlier medical training, in the late 1820s young Charles had switched his career path to become a prospective clergyman in the Church of England. Under the tutelage of Reverend Henslow at the University of Cambridge, Darwin read Paley's works including an 1826 edition of *Natural Theology* and was deeply influenced by them.[21] For example, Darwin later recalls in his autobiography that Paley's logic "gave me as much delight as did Euclid" and that it was the "part of the Academical Course which...was the most use to me in the education of my mind."

Darwin never became a cleric, but the training left its intellectual mark. Much later in his career, after the voyage of the Beagle, Darwin increasingly struggled to reconcile his growing understanding of evolutionary processes with fundamentalist teachings of the Church. Even then, however, he sometimes seemed reluctant to remove the paintbrush of evolution from a sentient deity's direct artistic hand. Indeed, Darwin retained a sense of the Almighty in his contemplations of nature and thereby also showed that evolution and deism are not necessarily incompatible (a sentiment that today remains quite common within as well as outside the halls of science). For example, in the second edition of *The Origin of Species*, Darwin mentioned God explicitly in the concluding paragraph: "There is grandeur in this view of life, with its several powers, having been originally breathed by the Creator into a few forms or into one."

Darwin's elucidation of natural selection was the most influential scholarly achievement in the history of science, fundamentally altering modes of thought and inquiry not only in biology but also in the social sciences and philosophy. Notwithstanding Darwin's use of "Creator" in the quotation above, this erstwhile cleric-in-training had discovered a mindless but nonetheless highly creative mechanistic force (natural selection) that apparently could craft spectacular and complex biological outcomes without direct supernatural intervention. Except for a few evolutionary biologists before Darwin (such as Jean-Baptiste Lamarck), and others who were contemporary with Darwin (notably Alfred

Russel Wallace, the co-discoverer of natural selection), most people in all prior times could only imagine that such biological outcomes had been the work of a sentient designer.

Darwin's theory of natural selection as the primary force of evolution was not adopted as "gospel truth" at the outset. Instead, it was given close scientific scrutiny, and it was not widely accepted in biology perhaps until early in the 20th century. This highlights one of the primary strengths of science, namely, that scientific theories are not accepted by force of authority but rather by force of evidence.

In obvious reference to Paley's timepiece metaphor, the modern scholar Richard Dawkins has described natural selection as a "blind watchmaker."[22] But natural selection is far less than blind. It is also an unconscious and amoral artisan, totally devoid of intelligence, foresight, and ethics. From among the multitudinous genetic variants that arise in each generation via the now well-understood hereditary processes of mutation and recombination, natural selection in effect makes choices about which genes survive and proliferate to populate each new generation. These unconscious decisions are based solely on genes' and organisms' reproductive performances (their relative genetic fitnesses), and they take into account neither a lineage's future prospects for evolutionary continuance nor an organism's personal well-being (except insofar as the latter may contribute to an individual's immediate genetic fitness).

Wherever genetic variation and differential reproduction occur (i.e., wherever life exists as we know it), natural selection transpires as an entirely mechanistic and inevitable process. Yet, most wondrous forms of life, humans included, have emerged from evolution's mindless machinations of heredity and natural selection. So too have emerged, however, multitudinous design flaws, as might be expected from unconscious as opposed to sentient and caring creative forces. Darwin's discovery of natural selection not only challenged natural theology's insistence on intelligence underlying biological design, but it also alleviated the age-old theological enigma of why sin and suffering exist in a world designed by a loving

God. Theodicic rationalizations no longer provided the only philosophical framework for explaining the age-old problem of evil.

Gregor Mendel, a scientific contemporary of Darwin during the mid-1800s, was the second of biology's two most important historical figures. Working with pea plants in a monastery in what is now the Czech Republic, this monk uncovered the particulate nature of hereditary factors (named "genes" in 1909). Mendel's key findings[23] about the nature of inheritance went unnoticed during his lifetime, but their rediscovery early in the 20th century eventually led to a successful merger of Darwinian and Mendelian principles that today remains the central platform (now vastly expanded and bolstered) of evolutionary biology. Mendel, the Augustinian friar, is another prime example of how, even well after the initial Age of Enlightenment, major scientific discoveries often have come from the devoutly religious.

In terms of epistemological significance, the Darwinian revolution of the 19th century (later wedded with Mendelian discoveries) essentially did for biology what the Copernican revolution of the 16th century had done for the physical sciences. Biology, like the physical sciences, was finally becoming unshackled from untestable religious doctrines. No longer, it seemed, was direct craftsmanship by a deity required to explain the marvelous behavioral, physiological, and morphological adaptations that enable organisms to survive and reproduce. No longer, it seemed, must *Homo sapiens* inevitably be viewed as a special product of divine creation or as being an inappropriate subject for objective inquiry in terms of natural shaping forces amenable to critical appraisal. No longer, it seemed, were nature's designs in general, and human makeups in particular, to be interpreted solely as the direct artisanship of almighty God.

Not everyone was convinced, of course.

THE INTELLIGENT DESIGN MOVEMENT

The Copernican and Darwinian Revolutions (plus other philosophical developments during Europe's Age of Enlightenment) had resulted in a diametrical shift in the fundamental relationship between religion and science. Prior to that time, a recurring issue was how to identify a secure place for science in a world dominated by religion. After then, a compelling issue (for some people) was how to identify a secure place for religion in a world seemingly dominated by science.

This is not to say that major religions (and certainly not their fundamentalist branches) have yielded the conceptual playing field to science. With respect to evolutionary issues, for example, a Creation Science (CS) movement in the United States continues today to make an oxymoron of itself by clinging tenaciously to a scientifically irrational belief (given voluminous countervailing evidence from physics, astronomy, biology, and geology) that all species were created in their current form in a universe that is only a few thousand years old. "Empirical support" for this hardline CS stance boils down to nothing more than a few short passages (Genesis 1:1–2:4), interpreted literally, from a Bronze-Age religious text. (The Bible is an amazing and powerful document, but it is clearly not a source of *scientific* knowledge—one can search its pages in vain for scientific insights on, for example, astrophysics, organic chemistry, genetics, or evolutionary biology.)

Creation Science would be unworthy of further discussion here except for the fact that in recent decades it has budded off a subsidiary movement, known as Intelligent Design (ID), that claims to have adduced scientific evidence that various biological traits were consciously (rather than unconsciously) engineered. Basically, the modern ID movement is Paley's natural theology reincarnate.

History

The modern ID version of natural theology can be dated to the publication in 1984 of *The Mystery of Life's Origin*, co-authored by Charles Thaxton (a chemist and historian), Walter Bradley (an engineer), and Roger Olsen (a geochemist).[24] This work had been organized and encouraged by Jon Buell, a former campus minister who became president of a Dallas-based Christian organization with an impressive title: the Foundation for Thought and Ethics (FTE). The book was intended to highlight difficulties with scientific explanations for how life on Earth might have arisen, and indeed it claimed categorically that life could not be explained by natural causes. A follow-up production sponsored by FTE was a high-school textbook (*Of Pandas and People*, published in 1989) by biologists Percival Davis and Dean Kenyon, which likewise took an unsympathetic stance on scientific evidence for evolution by natural causes. At least two states (Alabama and Idaho) approved this book for curricular use.

The ID movement got a much greater boost from the publication in 1991 of *Darwin on Trial*[25] by Phillip Johnson, a law professor at the University of California at Berkeley. The book garnered considerable media coverage and achieved popular success, probably in part because of Johnson's academic pedigree in a respected secular university but also because the author's arguments were generally more palatable to educated Christians than the scientifically outlandish claims often made by traditional creation scientists. In its antievolution stance, *Darwin on Trial* gave at least the aura of serious scholarship and did not overtly tout hard-line creationist mantras that were scientifically untenable (such as the young Earth scenario, or a universal flood).

In that same year, politician Bruce Chapman founded the Discovery Institute (DI), a conservative think tank based in Seattle, Washington. For the ensuing decade and continuing today, the DI has largely supplanted the FTE as the primary hub of activity for the Intelligent Design movement in North America. One branch of the DI—the Center for Science and Culture (CSC)—is devoted expressly to challenging Darwinian theory, promoting Intelligent Design, and exploring the impact of "scientific materialism" on human culture.

The next serious ID book—*Darwin's Black Box* by biochemist Michael Behe, 1996[26]—introduced the concept of "irreducible complexity." According to Behe, a cellular structure or any other biological feature is irreducibly complex if the removal of any one of its parts would result in a full loss of function. As such, the structure could not have evolved incrementally by natural selection, Behe claims, but instead must have been intelligently produced in its entirety for its specified role. Behe gave an example by analogy. A naïve but logically objective observer would doubtless conclude, correctly in this case, that a mousetrap was intelligently designed (rather than assembled piecemeal by natural selection) because removal of any of its five working parts (platform, hammer, spring, catch, and hold-down bar) would render the trap useless for its purpose of catching mice. Although the mousetrap example may have been ill chosen to illustrate Behe's concept of irreducible complexity (because critics quickly pointed out that functional mousetraps with fewer than five working parts can readily be envisioned),[27] the broader notion that particular biological features, such as a bacterium's flagellum (tail) or the human eye, might be irreducibly complex, became a rallying point for advocates of Intelligent Design.

Another landmark publication in the ID movement was William Dembski's *No Free Lunch: Why Specified Complexity Cannot Be Purchased without Intelligence*, published in 2001.[28] According to Dembski, who is a mathematician, a "design inference" is scientifically necessary whenever a complex phenomenon occurs with a small but specified probability, the specification coming from independent knowledge about the system. For example, if 10 arrows are found within the bull's-eyes of 10 targets, and if it is certain (i.e., specified) that the targets were in existence before the arrows were shot (rather than having been drawn a posteriori around previously shot arrows), then the logical inference is that the outcome cannot be explained either by natural laws or by chance and thus must be evidence of sentient skill (i.e., of intelligence). Dembski's arguments appear sophisticated because they are couched in terms of quantitative statistical theory (albeit laced with questionable assumptions), but in essence they merely echo qualitative CS sentiments against evolution based on a supposed improbability that any complex biological outcome could emerge from natural processes.

Courtroom Decisions

Throughout the 20th century and continuing today, biblical fundamentalists have promoted legislation that would oppose the teaching of evolution in biology classrooms. A spokesperson for some of the earliest antievolution efforts was William Jennings Bryan—a three times unsuccessful candidate for the U.S. presidency—who played a leading role in the famous Scopes "monkey trial" of 1925. John Scopes was a high school teacher in Dayton, Tennessee, who admitted violating a state law that forbade the teaching in public schools of "any theory that denies the story of the Divine Creation of man as taught in the Bible."[29] Three other states (Arkansas, Mississippi, and Oklahoma) likewise had passed laws prohibiting evolutionary instruction in public schools. The Tennessee court found Scopes guilty, but the well-publicized courtroom shenanigans probably tarnished more than abetted the creationist movement in the broader court of public opinion.

In the latter half of the 20th century, two decisions by the United States Supreme Court denied creationist challenges to evolution, in each case on the grounds that promoting religion in public schools is unconstitutional. In *Epperson v. Arkansas* (1968), the court concluded that creationism is religion, not science, and that

> Government in our democracy, state and national, must be neutral in matters of religious theory, doctrine, and practice. It may not be hostile to any religion or to the advocacy of non-religion, and it may not aid, foster, or promote one religion or religious theory against another or even against the militant opposite.

Similarly, in *Edwards v. Aguillard* (a 1987 case involving a Louisiana law that mandated the teaching of Creation Science together with evolution in public schools), the Supreme Court concluded that the

> primary purpose [of the Louisiana "Creation Act"] was to change the public school science curriculum to provide persuasive advantage to a particular religious doctrine that rejects the factual basis of evolution in its entirety. Thus, the Act is designed either to

> promote the theory of creation science that embodies a particular religious tenet or to prohibit the teaching of a scientific theory disfavored by certain religious sects. In either case, the Act violates the First Amendment.

Thus, the highest U.S. court has ruled consistently that Creation Science is a religious doctrine, not science. Implicit in these courtroom decisions was also the sentiment that if biblical creationism were to be discussed in classrooms supported by taxpayer dollars, then so too should be the creationist scenarios of other religions such as Hindu and Islam (as well as the views of mainstream Christianity and other major religions that can be more compatible with science).

The Intelligent Design movement has offered the most recent challenge to the teaching of evolution in public schools. In 2004, the Dover (Pennsylvania) Area School Board of Directors adopted a resolution that in effect emphasized perceived gaps in evolutionary theory and touted ID as a viable scientific alternative to be taught in public schools. Several parents objected, and this led to the several weeks long case of *Kitzmiller v. Dover Area School District* (2005) in the Federal District Court for the Middle District of Pennsylvania. In a 139-page decision, Federal Judge John E. Jones III reviewed the history of the creationist and ID movements in the United States and concluded the following:

> [W]e find that ID [Intelligent Design] is not science and cannot be adjudged a valid, accepted scientific theory, as it has failed to publish in peer-reviewed journals, engage in research and testing, and gain acceptance in the scientific community. ID, as noted, is grounded in theology, not science.... Moreover, ID's backers have sought to avoid the scientific scrutiny which we have now determined that it cannot withstand by advocating that the controversy, but not ID itself, should be taught in science class. This tactic is at best disingenuous, and at worst a canard. The goal of the IDM [Intelligent Design movement] is not to encourage critical thought, but to foment a revolution which would supplant evolutionary theory with ID.

Despite these courtroom setbacks, the ID movement continues.

Goals and Substance

Biology textbooks from ID authors are sometimes presented in quite scholarly style, and they also tend to depart from more traditional CS literature by seldom giving explicit mention to a Creator God or Christianity. The identity of the intelligent designer is often left vague or unspecified (because ID'ers are now well aware that the Constitution's First Amendment prohibits overt advocacy for any particular religious view in the science curricula of public schools). When pressed on the issue, proponents of ID may admit that God is the creative artisan, but the crux of the argument, they contend, is that many biological features give objective empirical evidence of intelligent rather than natural causation. The agent could be something other than God, such as extraterrestrials who may have seeded our planet with life (a suggestion made explicitly in ID's original treatise, *The Mystery of Life's Origin*). The agent could not, however, be natural selection, which is totally nonsentient.

The ID movement focuses on perceived weaknesses of natural processes in accounting for complex and well-functioning features of life. Thus, its criticisms of evolutionary biology tend to be negative rather than constructive (because appeals to supernatural causation are not scientifically helpful). Scientific facts, such as that the Earth is several billion years old or that dinosaurs were not contemporary with humans, are not necessarily disputed (Behe, for example, accepts that the universe is ancient),[30] so ID in some respects resembles the age-old "God-of-the-gaps" philosophy in which transcendent causation is invoked only when the scientific knowledge of the era is deemed inconclusive on a particular issue. This "argument from ignorance"[31] tends to make the ID movement inherently inimical to objective scientific inquiry and hostile to any findings that imply natural causation. In general, God-of-the-gaps notions tend to be rejected not only in biology (because they violate rules of scientific testability) but also in many theological circles (because they often seem to make God subject to diminution or irrelevance as scientific knowledge grows).

Proponents of ID are essentially united in their rejection of Darwinian explanations for life, and evolutionary biologists (as well as U.S. courtroom decisions) have been essentially united in their rejection of Intelligent Design as a serious scientific proposal. But neither group is monolithic and both do encompass a wide range of ideas. Evolutionary biologists collectively study and dispute competing *scientific* hypotheses regarding thousands of biological phenomena and their details; and the philosophical stances of the activists for ID can range from young-Earth creationism and species stasis to acknowledgments that the planet is billions of years old and that microevolution can occur within particular species. Some proponents of ID even go so far as to accept that similar species may once have shared common ancestors.

One interesting school of thought within ID believes that God front-loaded the universe at the time of the Big Bang and that all happenings since then (including the emergence and diversification of life) were ordained in God's original plan. Except for the preordained aspect of this unfolding, such views are somewhat reminiscent of the philosophical musings of Albert Einstein or Charles Darwin when these eminent scientists spoke in reverent terms about the geneses, respectively, of the physical universe and biodiversity. A relaxed version of this idea is that God engineered life indirectly rather than directly, simply by setting in motion natural laws and natural evolutionary processes from which life eventuated without further supernatural intervention.

Although the ID movement clearly is driven by religious or sociopolitical motivation, many of its practitioners (like their Creation Science predecessors) proclaim that their methods are scientific. Indeed, when the biochemist Michael Behe delves into the molecular workings of cells and finds astounding molecular complexity, he is at least implicitly adopting the scientific stance that empirical evidence (rather than doctrinal authoritarianism) is the proper foundation for drawing biological inferences. In this book, I will assume that readers, whether religious or not, are not disingenuous in their desire for objective evidence for or against the hypothesis that biological traits (including features of the human genome, in particular) were designed and built by a sentient agent, whatever that agent might be.

COMMON MISCONCEPTIONS ABOUT NATURAL EVOLUTIONARY PROCESSES

The crucial argument of Intelligent Design (much like that of natural theology in earlier centuries) is that life's beauties and complexities could not have arisen via natural causes, so a supreme intelligence must be involved. To "prove" that this is so, some ID proponents resort to gross distortions (or perhaps just display profound ignorance) about the evolutionary processes that they see as antithetical to religion and that they therefore seek to deny. What follows are several oft-perpetuated misconceptions about natural causation in biology and the science of evolutionary genetics.

Falsehood 1: Evolution Equals Unsophistication

Skeptics of evolution contend that unconscious natural forces are far too unsophisticated to account for the evolution of complex biological structures and functions. Whether this philosophical conjecture has merit is ultimately a matter for critical empirical appraisal. However, one general comment is relevant at the outset: the sophistication of many biological features is not in scientific dispute.

Indeed, science (not religion) has revealed how elaborate many biological features can be. For example, the human genome is now understood to be composed of more than three billion pairs of nucleotides (the building blocks of DNA) arranged, in part, into thousands of functional genes that encode RNA and protein molecules that interact in highly complicated metabolic pathways to perform multitudinous cellular functions. Biologists ranging from biochemists and geneticists to systematists and ecologists have uncovered complexities of nature that vastly surpass what theologians alone might ever have imagined, at structural and functional levels ranging from organic molecules and cells to tissues, organs, morphological and behavioral phenotypes, species, symbiotic associations, biological communities, and ecosystems. Thus, science has abundantly confirmed that the challenges of understanding biological design are indeed profound.

Falsehood 2: Evolution Equals Random Chance

Advocates of Intelligent Design contend that complex biological features cannot arise by chance, the implication being that chance equates to natural evolutionary processes and anti-chance equates to sentient forces. From a scientific vantage, however, the driving force of adaptive evolution—natural selection—is itself the antithesis of chance. Hereditary factors that promote organismal survival and reproduction in a particular environment tend to be precisely those that proliferate across the generations and thereby come to characterize natural populations. Whenever genetic variation and differential reproduction exist in nature (as they do in all known species), natural selection is inevitable, both logically and empirically. Biological traits that emerge from this inexorable operation may have the superficial aura of intelligent artistry, but that appearance is illusory (under a scientific interpretation). Natural selection can be a highly creative process (given a suitable supply of genetic variation to work from), but it is merely a mechanistic phenomenon—as inescapable and insentient as gravity.

This is not to say that evolution is devoid of important stochastic (i.e., chance) elements. Natural selection can sift only among the genetic variants available for its scrutiny, and two of the three primary sources of genetic variability—de novo mutation and recombination[32]—occur essentially at random with respect to forging adaptations.[33] The new mutations and recombinant genotypes that arise in each generation have no biased tendency to enhance either an individual's genetic fitness (its reproductive success relative to other individuals) or the adaptive needs of a species. In other words, favorable alleles and more fit genotypes have no known mutational tendency to arise disproportionately when needed. In this important sense, the genetic fodder upon which natural selection acts can indeed be characterized as stochastic or chancy in origin.[34]

The third source of population genetic variation entails a mixture of "chance and necessity."[35] Apart from de novo mutations and recombinant genotypes, the genetic variety available for natural selection in any generation is also a function of historical circumstance, that is, of idiosyncratic genealogical outcomes that have been affected by both stochastic and directive evolutionary processes across all prior generations. Evolution going forward can work only with the biological substrates provided by evolution foregone. These biological substrates—"ghosts of evolution past"—are not supernatural legacies, but instead they are real genetic lineages and real species that have been subjected for eons to the full panoply of evolutionary processes including natural selection (the directive agent of adaptive evolution) as well as idiosyncratic mutation, recombination, and genetic drift[36] (stochastic forces in the sense described above).

The temporal nature of heredity also means that evolution is inherently a phylogenetic process, involving descent with modification. So, for example, when two or more species share exquisite details in some complex biological feature (such as a long nucleotide sequence for a protein-coding gene), the usual evolutionary interpretation is that these species inherited copies of that trait from a shared ancestor. The creationist explanation, by contrast, posits that God created such traits independently in each species, starting in each case from scratch. At least at a superficial explanatory level, evolutionary and creationist scenarios both seem plausible, in principle, for complex traits that perform their functions well. A more acid test comes from complex traits that are harmful to their bearers. As we will see in later chapters, many complex genetic traits (such as pseudogenes and mobile elements) that often are functionless or even detrimental to the organisms that house them are rampant in the genomes of vertebrate animals, humans included. Did a Creator God repeat these apparent errors of genomic construction time and time again? Or are such genomic flaws merely the footprints of phylogenetic history?

Evolutionary processes do not contrive complexity directly from nothing. Natural evolutionary processes operate on genetic lineages much like homeowners work on their houses—by taking advantage of available construction materials to make individually small but sometimes cumulatively substantial alterations to previously existing structures and functions. On occasion, fairly extensive renovations may occur rather quickly,[37] but the norm—both for houses and biological lineages—is evolutionary gradualism in which renovated forms closely resemble their immediate predecessors and increasingly diverge from the preexisting entities after longer periods of time and impetus.[38] In the case of biology, fossils and molecular evidence indicate that most grand evolutionary transitions (such as from ancestral fishes to various derived groups of terrestrial vertebrates) require millions of years and involve many intermediate steps.

The analogy of natural selection to a homeowner can be carried too far, however. A homeowner presumably has a longer-term plan or intent for her renovations, whereas natural selection renovates biological features without foresight. The force of natural selection acts as if myopic, unable to perceive the longer-term consequences of its immediate decisions, which are based solely on the fit of available genotypes to current adaptive needs.

At any horizon in time, history-laden genetic lineages both facilitate and constrain what evolution might accomplish going forward. Natural selection is facilitated in the sense that it need not re-contrive complex biological features ex nihilo in each generation; but it is also constrained because it must operate within the framework of the available (phylo)genetic materials that underlie the existing biological structures and functions in each evolved lineage.[39] This raises a key point germane to this book. Evolutionary causation via natural processes leads to a biological expectation not shared by most versions of ID: a routine appearance of suboptimal organic design.

Falsehood 3: Natural Selection Ensures that Only Favorable Genes Establish

Whereas falsehood 2 perpetuates the incorrect notion that evolution by natural causes is entirely chancy and thus could never yield functionally complex biological traits, falsehood 3 perpetuates a diametrically opposite but equally incorrect notion that natural selection invariably ensures favorable biological outcomes. Natural selection can be powerful, but it is never all-powerful. Instead, it is just one in a nexus of evolutionary forces, others of which can override the adaptation-promoting power of natural selection in particular instances and thereby yield suboptimal or even detrimental biological outcomes. The following are examples of evolutionary genetic forces that routinely counter the efficacy of natural selection:

Genetic Drift

Especially in small populations, gene frequencies tend to change from generation to generation, even in the absence of natural selection, due to the chance sampling of gametes. This stochastic genetic process is analogous to what happens if a child were to draw a small handful of jellybeans from a jar containing two or more well-mixed types. Suppose that red and blue jellybeans are at equal frequency ($p = q = 0.5$) in a jar from which a child randomly grabs four beans. The statistical probability that the child drew two red jellybeans and two blue jellybeans is $6p^2q^2 = 6(0.5)^2(0.5)^2 = 0.375$; all other outcomes (with a collective probability in this case of $1 - 0.375 = 0.625$) would represent a stochastic change in the frequency of jellybeans in the sample compared to that in the original jar. The analogy, of course, is to different forms of a gene (i.e., different alleles), which like the jellybeans from a jar (analogous to a population's gene pool) are subject to random fluctuations in frequency when a finite number of gametes from the population are drawn to initiate each new generation of organisms.

SLIGHTLY DELETERIOUS ALLELES

Some deleterious genes are nearly neutral (i.e., only slightly harmful) and thus fail to register clearly on the radar screen of natural selection. This is especially true in small populations where stochastic effects of genetic drift become magnified, potentially overriding what would otherwise be more effective selection against the deleterious DNA. A gene that is only mildly deleterious may escape selective elimination from the host population for long periods of evolutionary time; and if many such genes accumulate in the genome, their collective genetic burden to individuals and to the population can be substantial.

GENETIC CORRELATIONS AND CONFLICTS

In many cases, a deleterious allele may be linked to or otherwise correlated in transmission with a host-beneficial allele at another gene. In such cases, the deleterious allele can hitchhike with the favorable allele and thereby be maintained in a population despite what otherwise might be the best efforts of natural selection to eradicate it.

PLEIOTROPY AND TRADE-OFFS

Pleiotropy is a phenomenon wherein a particular gene or set of genes has multiple phenotypic consequences for an organism. Fitness trade-offs can ensue if the resulting influence on one trait improves a person's well-being while the influence on another trait diminishes a person's health. For example, any genes that were under positive selection for increased skull and brain size (and better cognition) in human evolution also have contributed to the difficulties of childbirth, as infants' heads gradually became too large to pass easily through the available birth canals. In another such example, any genes predisposing for the calcification of body parts might be beneficial in terms of strengthening a person's bones and yet detrimental in terms of promoting atherosclerosis (calcification of artery walls). "Antagonistic pleiotropies" of this sort are no doubt common, and they provide yet another evolutionary reason that biotic features often reflect biological compromises rather than ideal outcomes.

SELFISH GENES

In sexual species including humans, natural selection operates not only via differential organismal reproduction but also via the differential proliferation of particular genes within a lineage. As we will see in later chapters, various pieces of DNA have discovered selfish means to multiply themselves within the human genome, sometimes to astonishing numbers. Such sequences in effect have profiteered by taking advantage of natural selection operating at the level of the gene, which can run counter to the operation of natural selection working at the level of organismal fitness. Such genic-level selection can also lead to ever-shifting genomic alliances among unlinked loci in sexual taxa, producing conflicts of interest among otherwise collaborative genes and opening windows of opportunity for the evolution of selfish or parasitic (in addition to cooperative) behaviors by specific pieces of DNA. Selfishness in this context means that an endogenous DNA sequence can sometimes persist and proliferate copies of itself in the genomes of a host population without enhancing either the immediate or long-term well-being of the hosts. Mobile elements (chapter 4) provide quintessential examples of selfish DNA sequences in sexual species.

SEXUAL SELECTION

Darwin made a clear distinction between traits that evolve under the influence of natural selection and those that evolve under sexual selection. The latter is the differential ability of individuals of the two genders to acquire mates, and the topic can be subdivided into two components: intrasexual selection, which refers to competition among members of the same sex over access to mates or to the gametes of the alternate sex; and intersexual or epigamic selection, which refers to mating choices made between males and females (or sometimes between male and female gametes). One reason for distinguishing sexual selection from natural selection is that only the latter normally promotes genuine *adaptations* to particular environments; sexual selection, by contrast, produces traits that are "adaptive" only to the demands of mating or gametic union, and that otherwise might be counterproductive to the individual. For example, some sexually selected traits such as the peacock's astonishing tail, although beneficial in attracting mates, is otherwise likely to be harmful by impeding flight and making males more visible and vulnerable to predators. Sexual selection is yet another source of often competing demands on genetic fitness, and yet another evolutionary reason that species routinely evolve biological features that do not fully conform to the hypothetical ideal.

HISTORY, OR COMMON DESCENT

I have already emphasized that at any point in geological time, evolution going forward can only work with the genetic diversity presented by lineages that have survived from the past. This too places severe constraints on what evolution by natural selection can accomplish. By contrast, a competing hypothesis—which is a hallmark of Creation Science—supposes that an intelligent agent can conjure (or at least has conjured in the past) de novo lifeforms at will.

All of these and other population genetic factors mean that natural selection (the primary agent responsible for biological adaptations) is not solely in charge of evolutionary outcomes. This reiterates the important point that evolution by unconscious natural processes (rather than by intelligent supernatural forces) correctly anticipates the routine appearance of suboptimal organic design. As this book will demonstrate, such suboptimality is pervasive in humans, too, not only at the traditionally observable level of phenotypes but also in the molecules ensconced deep within the human genome itself.

Falsehood 4: Complexity Equals Evolutionary Improbability

Another ID sentiment is that the emergence of biotic complexity via natural processes (unlike under intelligent auspices) is highly improbable, a notion that proponents of ID sometimes claim to document statistically. An undisputed law of statistics states that the probability of the joint occurrence of two or more *independent* events is calculated by multiplying together the separate probabilities of those events. For example, suppose that independent events A and B occur with random probabilities of 1 in 10 million (i.e, 10^{-7}) and 1 in 100 million (10^{-8}), respectively. The probability that both events occur together, by chance, is thus minuscule: one in a quadrillion (i.e., $10^{-7} \times 10^{-8} = 10^{-15}$). Mutations are examples of such rare and independent events, each typically occurring at a rate of about 10^{-7} or 10^{-8} per gene per generation. Suppose that a complex adaptation (such as a metabolic resistance to two different categories of otherwise lethal drugs) requires the joint presence in an organism of two mechanistically independent mutations. A proponent of ID might conclude that this adaptation is effectively unachievable by natural forces because its random probability is so very low. Divine intervention then becomes the default explanation.

But this is a gross misapplication of statistics, as the following example will illustrate. Consider a small colony of a billion bacterial cells, housed in a test tube, that lacks the genetic capacity to survive the antibiotics penicillin and streptomycin. New mutations for penicillin resistance (p^+) and streptomycin resistance (s^+) are known to arise randomly in bacteria, at low frequency. If the mutation rate to p^+ is a plausible 10^{-7}, then a culture of 10^9 bacteria should by chance contain about 100 cells with penicillin resistance. When penicillin is added to the culture, all bacteria die except for those lucky 100 cells which then quickly divide and multiply to bring their number back up to a billion cells (all now carrying the p^+ mutation). In this new colony, about 10 cells should by chance be resistant to streptomycin, assuming that the mutation rate to s^+ is a plausible 10^{-8}. If streptomycin is then added to the culture, all of the bacteria die except for these 10 cells, which again divide and multiply to repopulate the test tube with a billion descendants (all of which are now p^+ and s^+ jointly). By this stepwise process, the unconscious operation of selection has made virtually inevitable what might otherwise have been deemed impossible—the evolution of a complex adaptation.

This exercise is not merely academic. Real microbial species routinely evolve genetic resistance to the antibiotic compounds we employ widely to kill them. For example, penicillin was introduced in 1928 and microbes resistant to it first appeared in 1946; for streptomycin, the corresponding dates were 1943 and 1959. Several other widely disseminated antibiotics have had similar histories of drug development and the ensuing evolution of bacterial resistance.[40] In recent years, the evolutionary rate of microbial resistance actually has outpaced the rate of drug discovery, thus threatening an end to the medical era when antibiotics could be relied upon to treat bacterial infections. Especially troubling are "superbugs" that step by step have evolved genetic resistance to many of the medical profession's most commonly used antimicrobial drugs.

Biological complexity is consistent with the laws of probability, properly applied. The component parts of complex adaptations need not arise simultaneously (i.e., jointly), but instead can accumulate step by step through time. In sexual species, the relevant mutations can also occur in different places (i.e., different individuals) yet still be brought together through intermating. The only necessity in building a complex adaptation piecemeal by natural selection is that the intermediate stages must be adaptive (or at least not highly maladaptive).

Furthermore, the intermediate or penultimate stages of a complex adaptation need not perform the same function as the final edition. Evolutionary processes often yield traits whose present-day roles differ dramatically from those of their predecessors. A well-documented example at the level of morphology involves our inner-ear bones (mallus, incus, and stapes), which now are involved in our sense of hearing but that were part of the jaw apparatus in the ancient ancestors of mammals. An analogy to home renovation again applies, as for example when a garage might be converted to an office or spare bedroom. As will be described in later chapters, molecular biologists have uncovered numerous examples of evolutionary shifts in function for a wide variety of vertebrate genes and genomic features. Indeed, such shifts open windows of opportunity for stepwise evolution to include some very large steps, such as, for example, when mitochondria (power generating organelles inside a eukaryotic cell) emerged following the ancient evolutionary amalgamation of formerly independent microbes (chapter 3).

Proponents of intelligent design may contend that particular biological outcomes—such as bacterial resistance to multiple antibiotics—are irrelevant to broader arguments about the origins of biotic complexity when such traits are not irreducibly complex but rather can emerge step by step. In general, proponents of intelligent design would welcome any evidence that complex biotic features originated ex nihilo, whereas evolutionary biologists try to decipher exactly how complex adaptations have arisen stepwise by natural evolutionary forces. Judging from experience, debate between the two camps is not likely to be settled definitively in the court of public opinion so long as attention is confined to complex biological traits that function to near perfection; ergo, the emphasis in this book on complex biological traits that malfunction.

Falsehood 5: Natural Equals Proper

Concluding that a biological outcome arose by natural causes (in humans or in any other species) implies little or nothing about that outcome's ethical propriety in any deeper philosophical sense. As emphasized by the 18th-century philosopher David Hume, deducing "ought" from "is" represents a naturalistic fallacy because there is no scientific merit to the claim that what is biologically natural is necessarily morally proper (or, conversely, that biological naturalness is morally primitive or base). Science in this important sense is descriptive and dispassionate about the objective facts and mechanistic processes it uncovers. This is not to say that science is irrelevant to our understanding of ethics and religion. Indeed, I subscribe to the proposition that the biological sciences (and especially evolutionary biology) can hugely inform our attempts to answer typically religious questions, for example, about origins, fate, and meaning in life. In any event, anyone who might attack the field of evolution on the grounds that the science itself is inherently unethical or immoral is simply way off mark.

Falsehood 6: Evolutionism Necessitates Atheism

Although many professional evolutionary biologists are atheists or agnostics, many others are deeply religious or spiritual, and many are participating members of standard religious faiths, including Christianity.[41] Much depends on one's definition of God, or on one's perception of how God works (as evidenced by the quotation from Theodosius Dobzhansky presented earlier). Another reason that evolutionary biology is not necessarily equatable with atheism is epistemological: no science can prove a null hypothesis, which in the case of atheism is the notion that there is no God.

A PARADOX ETERNAL?

Experience suggests that no amount of scientific evidence or scientific argument will persuade fervent IDers that the beautiful designs and complex adaptations of life could have arisen by blind natural forces. Conversely, no amount of theological rhetoric seems likely to persuade those of scientific bent that complex adaptations are the direct result of supernatural causation. The impasse exists in part because religion and science have different operational and evidentiary paradigms. However, the impasse is exacerbated because Intelligent Design and natural selection *both* predict that biological systems (complex or otherwise) should operate effectively.

In this book, we will not butt our heads further against the wall that divides science and ID over explanations for instances of apparent biological perfection (or even for instances of mere functional adequacy). Instead, we will flip the issue on its head by focusing on overt biological imperfections and malfunctions. Does the evidence for sloppy craftsmanship extend into a molecular biological realm—the human genome—that until recently was simply beyond the empirical purview of scientists and theologians alike? If the answer is "yes," the theodicy challenge will again have reared its troublesome head, this time at the most elemental and potentially immortal level of our physical existence. If the answer is "no," biologists might have to reconsider a Platonian-like notion that idealistic perfection (in this case residing deep inside the human genome) materially underlies our outward biological imperfections. Indeed, if such molecular flawlessness truly exists, then science might even have to entertain the possibility that a supreme intelligence directly sponsored complex biological design.

THE HUMAN GENOME

The human genome that the research teams sequenced in 2001 was actually a composite of DNA sequences assembled from different individuals, but collectively it represented one "genomic equivalent" from our species. That representative genome was about three billion nucleotide pairs (or 3,000 megabase pairs) in total length. Thus, if each nucleotide pair in the human genome is considered analogous to one alphanumeric character in a written document, then the standard human genome is approximately 8,200 times longer than the book you currently are reading. The size and apparent complexity of the human genome are a source of wonderment to scientists and nonscientists alike, as this book will attest. Table 1.1 summarizes some of the major features of the human genome that will be discussed at depth in subsequent chapters.

TABLE 1.1 Various categories of DNA and their proportionate representation (expressed as a %) in the full human genome that consists of approximately 3,000,000,000 nucleotide pairs

Type of DNA	discussed in...	% of the human genome
intergenic DNA	Chapter 3	68.3
mobile genetic elements	Chapter 4	≥44.6
introns internal to protein-coding genes	Chapter 3	30.8
untranslated regions flanking genes	Chapter 3	5.1
microsatellites	Chapter 4	3.0
protein-coding sequences (exons)	Chapters 2, 3	0.8
pseudogenes	Chapter 4	0.7
mitochondrial DNA	Chapter 3	«0.1

* Percentages do not sum to 1.0 because various categories overlap. For example, much of intergenic DNA consists of mobile elements.

Source: Modified from table 3.1 in: Lynch, M., 2007, *The Origins of Genome Architecture*, Sinauer, Sunderland, MA.

Since 2001, DNA technologies have steadily improved and become streamlined such that whole-genome sequencing is now (or soon will become) almost routine.[42] In 2007, the full genome from a single human was sequenced in its entirety,[43] and similar reports quickly followed of whole-genome sequences from additional people.[44] Recently, a Personal Genome Project was announced,[45] the goal of which is to use rapid DNA-sequencing methods to gather genomic-level information from many individuals. No two human genomes are likely to be 100% identical in nucleotide sequence, so a compelling scientific challenge for the 21st century will be to screen genomes in ways that might enable doctors to personalize or tailor medical treatments and pharmaceuticals to the genetic peculiarities of particular patients and specific disorders.[46] Basic science will also be the beneficiary of such genomic screening, as scientists learn more and more about the full scope of human genomic operations.

In the meantime, geneticists and biochemists already have uncovered ample empirical evidence to address the central question of this book: Do the various structural and functional features that are well documented for human genomes insinuate intelligent design by a conscious engineer, or do they seem to register nonsentient craftsmanship by natural evolutionary forces?

AN INTRODUCTORY DISCLAIMER

Some readers might anticipate this book to be a frontal assault on religion or theology. It is not. Indeed, it is not an attack on any particular philosophy, except perhaps what I interpret to be an unacceptable stance—often promoted in the evangelical movements of Creation Science and Intelligent Design—that evolutionary science is irrelevant, misleading, or even damnable in its objective efforts to explore the human condition. Many scientists and philosophers see no necessary incompatibility between science (including the evolutionary sciences) and religion (including mainstream Christianity), either because these two approaches are perceived to address separate realms or magisteriai,[47]or because they need not disagree (and might even be interpreted as complementary) in addressing phenomena that reside in overlapping philosophical jurisdictions.[48]

I have three primary goals in this book: to help to educate a broad audience about the inner workings of the human genome; to challenge proponents of Intelligent Design to address, more critically, the ancient theodicy challenge as it applies at the biomolecular level; and in general to promote the evolutionary sciences as a preferred means to comprehend biological phenomena. Specifically, I hope to nurture the sentiment that objective science—in this case from the field of molecular evolutionary genetics—can inform humanity's attempts to understand the etiologies of complex biotic traits and processes.

CHAPTER 2

FALLIBLE DESIGN: PROTEIN-CODING DNA SEQUENCES

By the early 1900s, doctors had begun to appreciate that biochemical malfunctions inside the human body can produce physical ailments and abnormalities. Alkaptonuria is a rare genetic disorder (one in 200,000 births) that scientists now understand is caused by a defect in the metabolic pathway for phenylalanine and tyrosine (two of the 20 amino acids from which cells construct proteins). A biochemical foul-up makes cells unable to metabolize homogentisic acid, which therefore builds up and binds inappropriately to fibrous connective tissues in the patient's body. The medical symptoms of alkaptonuria are degenerative arthritis in the large joints and spine, often becoming pronounced by mid-life. Clinical diagnosis is based on a characteristic darkening of a patient's cartilage and an excess of homogentisic acid in urine (which turns black when exposed to air).

Sir Archibald Garrod pioneered scientific research on alkaptonuria at the turn of the 20th century. Born into a well-to-do physician's family in London in 1857, Garrod devoted his career to the study of human metabolic abnormalities. Garrod was unaware of genetic principles at the outset, but the inheritance patterns he observed in families affected by alkaptonuria soon offered preliminary evidence that humans display the same general types of hereditary mechanisms (such as segregation) that Mendel had discovered in pea plants a half-century earlier. Garrod studied several other heritable disorders, including albinism, cystinuria, and pentosauria, and in 1909 he published a compilation of human genetic diseases in a book entitled *Inborn Errors of Metabolism*.[1] In revealing an importance for hereditary factors, Garrod, like Mendel, was far ahead of his time: several of his observations came in 1901, which also happens to be almost exactly 50 years before DNA was confirmed to be the genetic material of life,[2] and 100 years before the nucleotide sequence of a human genome was ascertained.

Garrod is the father of biochemical genetics, but his legacy in a sense goes well beyond genetics and medicine. Garrod had introduced an unorthodox notion—that the metabolic machinery of the human body has inherent and oft-devastating design flaws. With the benefit of hindsight, we can see a philosophical consequence for these discoveries that Garrod himself either did not recognize or at least did not emphasize in his writings: an extension of the theodicy challenge into the previously unexplored realm of biochemistry. No longer could rational observers rightly invoke exogenous agents (such as poor air or disease-causing microbes) or malevolent supernatural forces (such as bad karma or demons) as the proximate source of all human ailments. Biologists in Garrod's time were just beginning to understand that heritable factors (genes)—endogenous to the human body—can affect the biochemical workings of cells, and that inborn errors of metabolism can profoundly undermine a person's health. The question we might now ask is, Why would an intelligent designer have crafted the innermost machinery of human life to be error prone? But before turning to that philosophical issue, we first must ask whether the genetic diseases uncovered by Garrod essentially exhausts the list, or were these disorders merely the tip of a vast iceberg of mutation-based genetic problems for *Homo sapiens*?

A BRIEF HISTORY OF BIOCHEMISTRY AND GENETICS

Biochemistry is the study of life's chemical processes, including the complex metabolic pathways by which living organisms construct (anabolize) and break down (catabolize) organic molecules. Genetics is the study of how heritable molecules are expressed within the individual and transmitted across the generations. In the century following Garrod, biochemists and geneticists gradually accumulated vast amounts of structural and functional information on thousands of proteins encoded by Mendelian genes.

Biochemists in the early 20th century devoted special attention to proteins, a huge class of organic molecules composed of long chains of amino acids. Proteins are structural and functional workhorses of cells. Some proteins make up the physical fabric of cells, tissues, and intercellular matrices. Actin, for example, is a primary constituent of microfilaments in muscle cells, and collagen is a major component of ligaments, cartilage, and tendons. Other proteins are enzymes: organic catalysts that speed up the rates of biochemical reactions, enabling oft-complex metabolic networks to operate effectively.

By the mid-1900s, sufficient information from detailed biochemical detective work had allowed scientists to decipher several metabolic pathways shared by humans and other animals. For example, in 1937 Hans Krebs correctly postulated the mechanism of the tricarboxylic acid (TCA) cycle (also known as the Krebs cycle or the citric acid cycle) wherein pyruvic acid is oxidized to carbon dioxide and water, generating in the process some of the chemical energy that a cell needs to survive and function (figure 2.1).[3] This effort earned Krebs the Nobel Prize for medicine in 1953. In the TCA cycle, a series of biochemical reactions converts pyruvic acid successively to acetyl-CoA, citric acid, isocitric acid, α-ketoglutaric acid, succinyl-CoA, succinic acid, fumaric acid, malic acid, and oxaloacetic acid. Each step in the pathway is facilitated (i.e., catalyzed) by a specific enzyme: pyruvic dehydro-

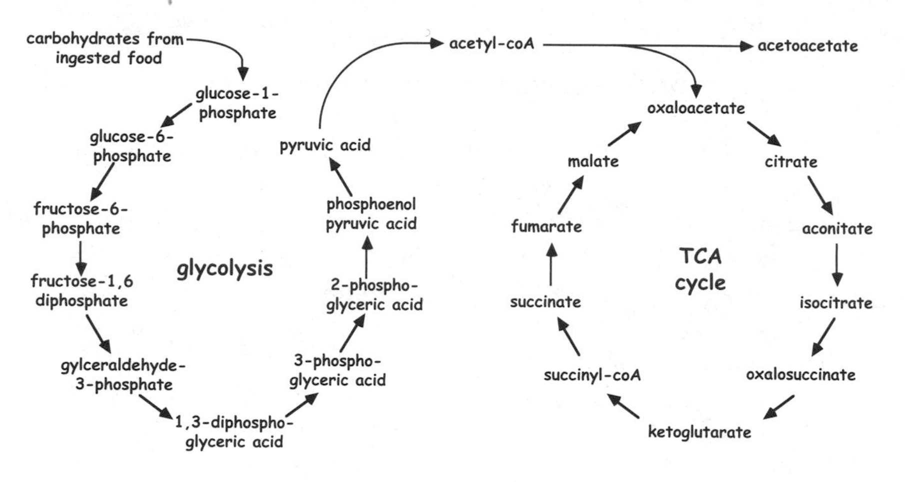

FIGURE 2.1. The complexity of biochemical pathways as illustrated by glycolysis and the TCA cycle (just two among more than 80 metabolic pathways that are well characterized in humans).

genase, citrate synthase, aconitase, isocitrate dehydrogenase, α-ketoglutarate dehydrogenase, succinyl thiokinase, succinic dehydrogenase, fumarase, and malic dehydrogenase. The pyruvic acid itself (a starting point for the TCA cycle) is the end product of another lengthy metabolic chain—glycolysis (figure 2.1)—that typically begins with glucose (a constituent of the carbohydrates that we ingest as food).

In the ensuing decades, biochemists gradually elucidated operational details in dozens of metabolic pathways within the bustling biochemical factories that we call cells. These pathways interconvert, shuttle, or otherwise manipulate various organic molecules of diverse classes including lipids, proteins, vitamins and cofactors, simple and complex carbohydrates, and nucleic acids as well as constituent parts and secondary compounds from all of the above. In a typically anabolic or catabolic pathway, each successive biochemical step either adds a little bit to or subtracts a little bit from the molecule that was the substrate for the reaction, much as a child might add or subtract different pieces of an erector set to gradually build or disassemble a complex structure.

Other metabolic pathways involve transporting particular compounds where they need to go, such as across nuclear membranes, or into or out of particular cellular organelles such as mitochondria (see chapter 3), lysosomes (intracellular vesicles that serve as constituents of intracellular digestion), or peroxisomes (which are sites of hydrogen peroxide metabolism).

The number of biochemical pathways in humans is hard to specify precisely. The difficulty stems in part from incomplete information and the complexity of metabolism, but it also reflects the fact that giving pathway designation to a set of biochemical reactions can be rather arbitrary. Adjacent pathways are structurally and functionally interconnected to varying degrees, so exactly where to draw boundaries between them is often unclear. In any event, one of the most robust attempts to reconstruct the full human "metabolic landscape" was published in 2007.[4] Based on an analysis of 2,766 documented metabolites and 3,311 metabolic and transport reactions in human cells, the authors enumerated a total of more than 80 metabolic pathways, including the TCA and glycolytic cycles. The list is probably far from complete, but whatever the final tally of biochemical pathways in the human body may prove eventually to be, the number of documented biochemical steps already is huge. So too, therefore, is the total number of trip-up points where biochemical missteps might occur. And occur they routinely do—often with dreadful health consequences—as we shall later see.

Before 1950, it was widely suspected that proteins would prove to be the genetic material of life. The complexity of organisms seemed to demand a comparable complexity for hereditary factors; and proteins—typically being composed of lengthy stretches of as many as 20 different types of amino acid—were known to be structurally elaborate and functionally diverse. DNA molecules, by contrast, seemed unlikely candidates for genetic material because nucleic acids are composed of only four types of building blocks: the nucleotides adenine (A), guanine (G), thymine (T), and cytosine (C). Thus, the scientific world was astonished to learn, beginning in 1944, that DNA—a supposedly boring and monotonous molecule—is the hereditary stuff of life.[5] It turns out that DNA's vast scope for encoding molecular complexity and biochemical diversity derives from the multitudinous sequence combinations possible when even small numbers of different types of nucleotide subunits (A, G, T, and C) are strung together in long chains. An obvious analogy is to the Morse code, wherein merely two types of subunits (dots and dashes) can be joined in endless sequence combinations to convey all words, sentences, and paragraphs in the human language.

Also in the mid-1900s, scientists began to uncover details about the special biochemical relationship between nucleic acids and proteins. In 1941, George Beadle and Edward Tatum[6] proposed, from their research findings on the fungus *Neurospora crassa,* that one gene seems to equate to one protein, and scientists soon worked out why this is so. Protein production from DNA is basically a two-step process: transcription and translation. During transcription, which takes place in a cell's nucleus (where chromosomes are housed), the linear sequence of deoxyribo-nucleotides in a particular gene is copied to a complementary linear sequence of ribo-nucleotides in messenger (m) RNA. The mRNA then moves outside the nucleus to ribosomes in a cell's cytoplasm, where its sequence of nucleotides is converted (translated) to the chain of amino acids that makes up a particular polypeptide (a string of amino acids that becomes part or all of a protein). The overall flow of information is thus DNA→RNA→protein.[7] The sequence of nucleotides in a particular polypeptide-coding gene ultimately dictates the sequence of amino acids in the enzyme or other protein for which that DNA is responsible (ergo the famous phrase: one gene, one polypeptide).[8] A remarkable fact is that transcription and translation (like many other basic molecular processes in genetics) are essentially universal to life, occurring similarly in creatures otherwise as different as plants, fungi, and humans.

A key step along the historical path to this scientific understanding involved cracking the genetic code. During the translation process, transfer (t) RNA molecules transport free amino acids to ribosomes where the mRNA molecules are translated into proteins. Each type of tRNA carries a particular type of amino acid (such as valine or methionine), which it transfers when called upon to the growing end of a polypeptide chain. The code was cracked when biochemists determined that each triplet of nucleotides in mRNA specifies one-and-only-one amino acid, and when they then figured out precisely which amino acid each triplet dictates.[9] For example, the triplet GGA (guanine-guanine-adenine) in mRNA commands an appropriate tRNA to add the amino acid glycine to a growing polypeptide chain, and the triplet ACG (adenine-cytosine-guanine) specifies an addition of the amino acid threonine. Altogether there are $4^3 = 64$ different triplet sequences possible with four types of nucleotides, but only 20 different types of amino acids. This means that the genetic code is "redundant": different triplets in a nucleic acid sometimes specify the same amino acid (e.g., AAA and AAG both translate to lysine). Each triplet specifies only one amino acid, however, so a given nucleotide sequence in a protein-coding gene dictates just one specific polypeptide chain. Finally, three "stop codons" do not specify any amino acid at all but instead act as signals for termination of the translation process.

All of this might be a functionally superb state of affairs for organisms except that random mutations (including point mutations that occur at particular nucleotide sites) arise routinely in protein-coding genes. Any nucleotide substitution that results in an altered amino acid sequence is a nonsynonymous mutation, whereas any substitution that fails to alter a polypeptide's structure—because of the genetic code's redundancy—is by definition a synonymous change. Based on the genetic code itself, approximately 75% of de novo random point substitutions in protein-coding DNA are expected to be nonsynonymous. Other random mutations involve short additions or deletions of nucleotides in a gene's coding region. When these alterations result in an improper shift in the reading frame by which successive triplets of nucleotides are transcribed and translated into a polypeptide, the mutation is called a frameshift. Most frameshift mutations, and some nonsynonymous mutations, substantially alter or destroy a protein's function, often with devastating health consequences.

By illuminating direct mechanistic links—transcription and translation—between genes (DNA) and proteins, biochemists had unveiled critical components in the proximate etiologies of inborn errors of metabolism. Namely, particular mutations in a protein-coding gene direct cells to produce a physically altered protein that may function poorly when called upon to contribute to a metabolic pathway; and a functional flaw in a metabolic pathway can produce serious health problems.[10]

BAD-HOUSEKEEPING GENES

Current estimates are that the human genome harbors about 24,000 protein-coding genes. These are sometimes called "good-housekeeping" loci because the enzymes and other proteins they encode busily conduct the myriad day-to-day metabolic chores in each cellular household. Most biochemical tasks (such as those illustrated in figure 2.1) are crucial for proper cellular structure and function. Thus, human geneticists (beginning with Garrod) initially were shocked to discover that members of this intracellular molecular workforce are disabled routinely by deleterious mutations. As the list of scientifically documented genetic disorders in humans grew dramatically during the 20th century, a new sentiment emerged: that particular mutations in almost any housekeeping gene could be harmful, with health consequences ranging from mild to deadly. Today, vast numbers of human genetic disorders have been illuminated at the molecular level, often in great detail. What sorts of things transpire when good-housekeeping genes are hit by random deleterious mutations?

When Protein-Coding Genes Go Bad

Alkaptonuria is one of the inborn errors of human metabolism studied by Garrod more than a century ago. In recent years, geneticists have discovered that the responsible gene (which encodes homogentisic acid oxidase, or *HGO*) is located on the "q" arm of chromosome 3, that it extends across 60,000 nucleotide-pairs of DNA, and that it consists of 14 polypeptide-specifying sections of DNA (i.e., exons; see chapter 3) interspersed with much longer stretches of noncoding DNA (known as introns). Each of at least 18 separate mutations, distributed across 11 of the exons, has been documented to disrupt *HGO*'s normal function so severely that homogentisic acid accumulates in the patient's tissues and produces serious clinical symptoms of alkaptonuria.[11] One of these point mutations leads to the substitution of one amino acid (adenine) for another (guanine) in exon 8; another is the insertion of a guanine in exon 7. Each mutation that underlies alkaptonuria behaves as a recessive allele, meaning that a person must inherit two copies of the defective gene (one from the mother and the other from the father) to express the disorder.[12] This metabolic abnormality typically is present from birth (as evidenced by diapers stained black by the urine of afflicted babies) and continues mercilessly throughout life.

Another genetic disorder is phenylketonuria (PKU), an enzyme deficiency that can cause mental retardation. The gene—for phenylalanine hydroxylase—resides on a distal arm of chromosome 12, and at that locus several dozen different recessive alleles (mostly due to nonsynonymous mutations that by definition lead to amino acid substitutions) have been found to underlie various cases of PKU in the human population. The biochemical basis of learning disabilities in PKU patients is uncertain but at least in some cases seems to be related to the reduced production of neurotransmitters in the brain. PKU can be treated by dietary means when diagnosed early, so it provides a transparent example of a genetic disorder that actually results (as do all genetic disorders, ultimately) from a functional "mismatch" between particular genes and particular environmental circumstances.

Most inborn metabolic disorders in humans show recessive inheritance. For the gene in question, this means that the normal or wild-type allele in a heterozygous individual is able to compensate for the dysfunctional mutant allele to the extent that the metabolic pathway is not seriously compromised and clinical symptoms fail to emerge. A few inborn metabolic disorders show dominant inheritance, meaning that even one copy of the deleterious mutation produces a clinical abnormality. One example is achondroplasia (short-limbed dwarfism), which results from a dominant mutation in the gene for a fibroblast growth factor receptor (FGFR). Another example is provided by erythropoietic protoporphyria (EPP), in which dominant mutations in the ferrochelatase gene hamper a person's ability to synthesize heme molecules that otherwise deliver oxygen to the body via the bloodstream.

Some genetic disorders in humans are sex-linked, meaning that the responsible gene resides on one of the sex chromosomes (X or Y) rather than on an autosome. In humans as in other mammals, each female normally has two copies of the X-chromosome (she is XX) whereas each male typically has only one copy (he is XY). Interestingly, the Y-chromosome is nearly devoid of functional genetic loci (apart from the male sex-determining gene itself), but the X-chromosome is loaded with them. The X-chromosome in the human species is also loaded with mutational defects, and these tend to cause extensive and disproportionate harm to males. One example involves X-linked hemophilia, made famous by its historical prevalence in European royal families (descendants of Queen Victoria [1819–1901]), where almost all of the afflicted were males. This sex bias arises because a single defective copy of an X-chromosome gene in an XY male normally is sufficient to produce this blood disorder, whereas the defective allele (which is recessive) in females is usually masked in expression by its wild-type counterpart. Other X-linked diseases in humans include particular forms of colorblindness, gout, Duchenne muscular dystrophy, Lesch-Nyhan syndrome, vitamin-D-resistant rickets, and *G6PD* deficiency.

Some genetic abnormalities are remarkably frequent in human populations. Glucose-6-phosphate dehydrogenase (*G6PD*) deficiency is the most common enzymopathy (enzyme-based disorder) known to science, affecting an estimated 400 million people worldwide.[13] In its normal functional state, *G6PD* catalyzes the initial step in an alternative route for glucose catabolism (separate from the glycolytic pathway) known as the pentose phosphate shunt. The enzyme is 515 amino acids long and is encoded by a gene whose 13 exons span a region of about 18,000 nucleotide positions on the long arm of the X chromosome. *G6PD* deficiency can result from any of approximately 130 different point mutations identified to date in various human populations, with the highest incidence rates (5–20%) being in malaria-infested regions of tropical Africa, subtropical and tropical Asia, the Middle East, parts of the Mediterranean region, and Papua New Guinea. Many people with *G6PD* deficiency remain symptom free, but clinical disabilities (including life-threatening neonatal jaundice and hemolytic anemia) can be triggered by exposure to particular drugs or infections, or by ingesting specific foods (notably fava beans). These symptoms may relate to the observation that blood cells in *G6PD*-deficient individuals seem prone to oxidative damage triggered by such environmental agents.

Many metabolic disorders involve nonenzymatic proteins. One example is sickle cell disease, wherein an abnormal form of hemoglobin (the oxygen-transporting molecule of blood) predisposes red blood cells to assume a rigid configuration, clog blood capillaries, and produce a painful and life-threatening condition in affected patients. Sickle cell disease is just one of many hemoglobinopathies (disorders related to hemoglobin) that involve structural alterations in one or another polypeptide subunit of the hemoglobin molecule.[14] Thalassemias constitute another category of hemoglobinopathies, in this case resulting from metabolic disruptions in the rate of production (rather than the biochemical structure per se) of hemoglobin molecules. Altogether, about 750 genetic variants for hemoglobin structure have been identified in human populations, and another 170 different mutations have been identified as the causal genetic basis for various thalassemias. Hemoglobinopathies are among the most common "single-gene" disorders in humans, with at least 5% of the world's population being genetic carriers for this category of inherited disease, and at least 370,000 severely affected individuals born each year.

Cystic fibrosis (CF) is a relatively common genetic disorder (about one in 2,500 births) attributable to deleterious mutations in a gene that encodes a nonenzymatic protein. This protein—cystic fibrosis transmembrane conductance regulator, or *CFTR*—is a key structural part of the membrane apparatus wherein microscopic pores or channels regulate the uptake of salt by cells. The *CFTR* gene contains 27 exons spanning about 250,000 nucleotide positions on the "q" arm of chromosome 7, and the protein it encodes is 1,480 amino acids long. More than 800 different CF-causing mutations have been uncovered, the vast majority of which involve substitutions or deletions of just one or a few nucleotides.[15] In Caucasian populations, the most common of these mutations produces a deletion of phenylalanine at position 508 in the *CFTR* polypeptide, which in turn precipitates a serious malfunction in ionic transport across the cell membrane. CF is a debilitating and often lethal genetic disorder that can affect many organs including pancreas and lung. It is a potent reminder of the devastation that sometimes can result from changing even a single critical nucleotide among the three-billion-plus nucleotide positions that comprise a human genome.

One Valiant Quest for a Gene

How do geneticists discover the precise molecular basis of a human biochemical disorder? The approaches are varied, but they typically entail DNA-level laboratory analyses interpreted in conjunction with medical and epidemiological evidence. To illustrate both the complexity and the human drama of the challenge, consider the remarkable research quest for the gene responsible for Huntington disease (HD), a fatal neurological disorder whose symptoms, usually beginning in mid-life, involve uncontrollable movements of the body and progressive dementia. In the United States alone, more than 25,000 patients suffer from HD, with about 125,000 more at risk by virtue of being siblings or children of the currently afflicted. The coarse-focus heredity of HD has long been known. By examining the pedigrees of affected families and interpreting the distribution of HD in the context of universal patterns of Mendelian inheritance, scientists deduced that a single gene must be involved, one defective copy of which inherited by either sex from either parent is sufficient to confer the disease on any person who lives long enough.

HD is most common in western Europe, but it also crops up in other hotspots such as Tasmania and Papua New Guinea. For Tasmania, the gene for HD can be traced to a widow who in 1848 left her home in England and moved with her 13 children to Australia. By 1964, descendants of this family accounted for most of the 120 afflicted people on the island. In Papua New Guinea, HD probably was introduced centuries ago by whalers from New England. Diaries tell of shipboard visits by "naked and friendly natives," some of whose children must have inherited copies of the HD gene from their sailor fathers.

In the whole world, the highest concentration of HD occurs in villages near Lake Maracaibo in Venezuela. The disease was introduced there (probably by a British sailor) in the early 19th century and eventually rose in frequency to more than 70 times the western European norm. At Lake Maracaibo, doctors have ministered to the sick, interviewed families, reconstructed pedigrees, and obtained blood samples for molecular analysis, all in a concerted effort to understand the finer genetic details of HD. Dr. Nancy Wexler played a marquee role in this drama. Originally trained as a clinical psychologist, Dr. Wexler decided to devote her life to the study of hereditary diseases when her own mother was diagnosed with Huntington disease. At Lake Maracaibo, she and her research team took advantage of the large family pedigrees (HD-affected families with a dozen or more children were common) and the availability of many polymorphic DNA markers whose locations on various human chromosomes were determined using molecular techniques and somatic cell hybridization. The latter method is little short of bizarre. When human and mouse cells are mixed in a test tube, they often fuse spontaneously into hybrid cells that contain an initial full complement of chromosomes from both species. Through successive divisions of these hybrid cells, the human chromosomes tend to be lost more or less at random, sometimes until only one remains. By matching the presence versus absence of particular human genes (as determined by molecular genetic assays) against the retention versus loss of human chromosomes in a panel of mouse/human cell lines, Wexler's team deduced which genes were housed on which chromosomes. The next step was to search for marker genes that co-segregated with the HD disease through the family pedigrees at Lake Maracaibo. By this correlational approach, the HD gene was localized to chromosome 4. Further molecular analyses (by laboratory techniques too detailed for description here) soon honed HD's position to the distal end of the short arm of that chromosome, and eventually to a specific address on that short arm.

Research efforts also have identified the precise structural abnormalities of the HD gene from which the metabolic malfunctions and debilitating symptoms arise. The HD gene (which encodes a cytoplasmic protein known as Htt) contains a terminal motif of nucleotide triplets (CAG, CAG, CAG,... CAG), the numbers of which are correlated with the expression of Huntington disease. On normal copies of the human fourth chromosome, 10 to 30 tandem copies of CAG are found in this "microsatellite" region, but more than 35 copies are present in HD sufferers. As described in chapter 4, Huntington disease is one on a growing list of hereditary disorders involving structural genes that display similar kinds of anomalies in the repeat motifs of microsatellites.

The valiant search for the HD gene has also led to a predictive diagnostic test (but as yet no cure) for the disease. This diagnostic test, which involves a laboratory count of the number of CAG repeats, illustrates an ethical quandary that Nancy Wexler refers to as the dilemma of Tiresias. In *Oedipus the King* by the Athenian philosopher Sophocles, the blind seer Tiresias confronts Oedipus with the thought: "It is but sorrow to be wise when wisdom profits not." In the current context, the dilemma of Tiresias can be phrased as a query: If one of your parents or family members has HD, would you wish to know whether you yourself are destined in midlife to contract this horrible and untreatable brain disorder? The availability of a diagnostic test means that a person can now gain such knowledge about his or her future; but to what avail? Thousands of individuals (including Dr. Wexler herself) have had to wrestle with this agonizing question. Many people have declined to take the diagnostic test, but others at risk for HD have chosen to learn their genetic fates.

Morbid Anatomy of the Human Genome

A landmark achievement occurred in 1960 with the publication of *The Metabolic Basis of Inherited Disease* (*MBID*),[16] which is the modern analogue of Garrod's *Inborn Errors of Metabolism* from half a century earlier. *MBID* has gone through more than half a dozen printed editions (at roughly six-year intervals), each a significant update and expansion of its predecessor. The title of the work also expanded—to *The Metabolic and Molecular Bases of Inherited Disease* (*MMBID*)—reflecting the recent explosion of DNA and protein sequence data that are providing unprecedented insight into human genetic disorders. The latest edition of *MMBID* occupies more than 6,000 information-packed pages in four weighty volumes. In 255 chapters, each on a different heritable disorder or suite of associated genetic disorders, leading biomedical experts encapsulate current knowledge about the molecular mechanisms underlying inborn human diseases. Attention is focused on the genetic basis of each disorder and also on proteomics (the structure and function of a gene's protein product). More than 500 well-characterized genetic disorders are profiled in astonishing detail, and that number will only grow—probably dramatically—as the medical profession moves more deeply into the modern era of intense genomic research.

The mutational defects profiled in *MMBID* occur in almost every operational category of protein, including enzyme-mediated energy metabolism, DNA/RNA processing, protein folding and degradation, molecular transport and secretion, signal transduction (mechanisms that link mechanical or chemical stimuli to cellular response), cytoskeletal elements, ribosomal functions, and structures and functions of exported (extracellular) proteins. With regard to affected tissues or organs, nothing appears sacred; various mutations are known to debilitate the nervous system, liver, pancreas, bones, eyes, ears, skin, urinary and reproductive tracts, the endocrine system, blood and other features of the circulatory system, muscles, joints, dentition, the immune system, digestive tract, limbs, lungs, and almost any other body part you can name. With respect to age of onset, various genetic disorders may appear in utero, birth to first year (the most commonly diagnosed class), year one to puberty, puberty to 50 years of age, or in seniors. Approximately two-thirds of the genetic defects discussed in *MMBID* shorten the human life span, and about three-quarters of these cause death before age 30.

Another compendium of this sort is *Mendelian Inheritance in Man* (*MIM*),[17] which describes thousands of human genes of which more than 75% have been reported to carry mutational defects associated with a disease condition (some examples are presented in table 2.1). *MIM* has appeared in a dozen printed editions and now is also available online (*OMIM*) where it is updated regularly by computer-based literature searches. When I checked on March 11, 2008, *OMIM* contained entries for 18,526 genes. Also published regularly as a part of MIM is a chart—a "morbid anatomy of the human genome"—showing the chromosomal locations of particular genes to which human disease-causing mutations have been mapped. One such chromosome from this morbid map is shown in figure 2.2.

One example is glycogen storage disease (*MIM* catalog number 232200), which involves a defect in the gene encoding glucose-6-phosphatase (an early step in glycolysis). Included in the *OMIM* entry are detailed descriptions of the chronology of scientific discoveries about glycogen storage disease (also known as Von Gierke disease), presented as synopses for a succession of

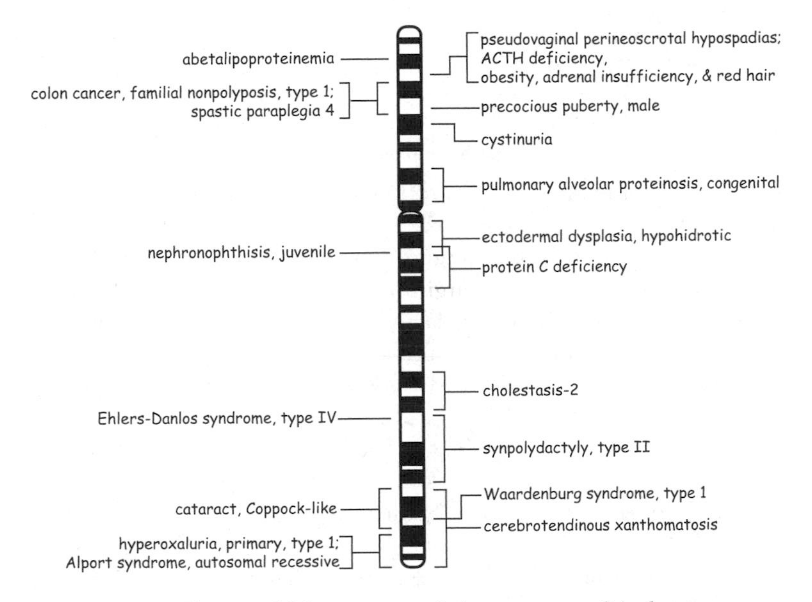

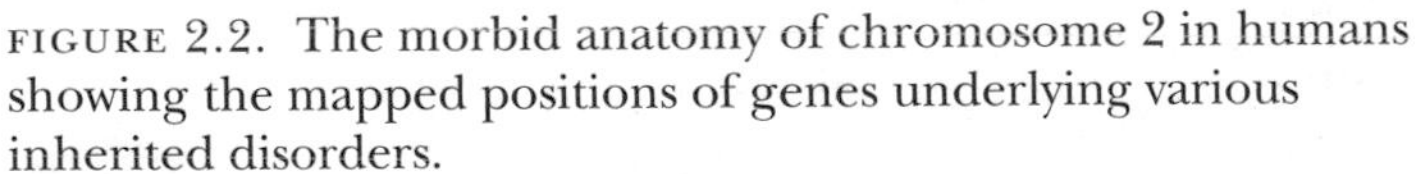

FIGURE 2.2. The morbid anatomy of chromosome 2 in humans showing the mapped positions of genes underlying various inherited disorders.

more than 40 scientific papers beginning in 1968; animal models for scientific research into the disorder; 14 distinct allelic variants (mutations) of the gene that are known to produce clinical symptoms of the disease in humans; and citations and computer links to about 70 relevant scientific publications from the primary research literature.

Another compilation of gene lesions responsible for inherited diseases is the Web-based *Human Gene Mutation Database* (*HGMD*).[18] Recent versions of *HGMD* describe more than 75,000 different disease-causing mutations identified to date in *Homo sapiens.* More than 50% of these molecular damages involve nonsynonymous substitutions in protein-coding segments (exons) of genes, whereas the remainder fall into a miscellany of categories including large and small insertions or deletions of genetic material, DNA rearrangements, regulatory mutations in regions that flank

TABLE 2.1 Some inborn metabolic disorders with relatively simple genetic etiologies in humans

Metabolic disorder	Enzyme usually showing mutational defect	*MIM* cat. #
acute intermittent porphyria	porphobilinogen deaminase	176000
argininosuccinate aciduria	argininosuccinate lyase	207900
citrullinaemia	argininosuccinate synthase	215700
classical galactosaemia	galactose-1-phosphate uridyl transferase	606999
CPS deficiency	carbamoyl phosphate synthase	237300
GABA transferase deficiency	4-aminobutyrate aminotransferase	137150
galactose kinase deficiency	galactokinase	230200
Gaucher disease	acid-β-glucosidase	230800
GSD type I Von Gierke disease	glucose-6-phosphatase	232200
GSD type III Cori disease	amylo-1,6-glucosidase or glucanotransferase	232400
GSD type V McArdle disease	muscle glycogen phosphorylase	232600
homocystinuria	cystathionine synthetase	236200
4-hydroxybutyric aciduria	succinic semialdehyde dehydrogenase	271980
hyperoxaluria type 1	alanine-glyoxylate aminotransferase	259900
isovaleric acidaemia	isovaleryl-CoA dehydrogenase	243500
ketone synthesis defect	hydroxymethylglutaryl CoA lyase	246450
lactose intolerance	lactase	223000
Lesch-Nyhan syndrome	hypoxanthine guanine phospho-ribosyl transferase	308000
maple syrup urine disease	branched chain ketoacid decarboxylase	248600
MCAD deficiency	medium chain acyl CoA dehydrogenase	607008

megaloblastic anaemia	dihydrofolate reductase	126060
methylmalonic acidaemia	methylmalonyl-CoA mutase	251000
Niemann-Pick disease	sphingomyelinase	257200
non-ketotic hyperglycinaemia	glycine dehydrogenase	238300
OCT deficiency	ornithine carbamoyl transferase	300161
phenylketonuria	phenylananine hydroxylase	261600
porphyria cutanea tarda	uroporphyrinogen decarboxylase	176100
proprionic acidaemia	propionyl-CoA carboxylase	606050
severe combined immunodeficiency	adenosine deaminase	102700
sucrose-isomaltose malabsorption	sucrase-isomaltase	609845
Tay Sach's disease	hexosaminidase A	272800
type V McArdle disease	muscle glycogen phosphorylase	232600
tyrosinaemia type I	fumarylacetoacetase	276700
tyrosine negative albinism	tyrosinase	203100
variegate porphyria	protoporphyrinogen oxidase	176200
xanthinuria	xanthine oxidase	278300

* Catalogue number from *Mendelian Inheritance in Man* (see text)

a gene, and alterations with consequences for how particular mRNA molecules are spliced together (see chapter 3). *HGMD* is cross-referenced to *OMIM* and updated weekly.

HGMD provides a concise summary of all documented mutations that underlie a given metabolic disorder. For example, typing "232200" (the *MIM* catalogue number for glycogen storage disease) into the *HGMD* Web site quickly brings up a screen indicating that 86 different disorder-causing mutations have been reported in the glucose-6-phosphatase gene. Sixty-three of these mutations are nonsynonymous substitutions, 18 are small additions and/or deletions, four involve splicing anomalies (see chapter 3), and one involves a regulatory region flanking the coding sequence. Another click of the mouse pulls up a Web page that describes additional molecular details of each mutation, from which we find that various cases of glycogen storage disease are known to be caused by a nucleotide substitution (T to G) in codon 5 of the glucose-6-phosphatase gene, which results in a switch from arginine to methionine at amino acid position 5 in the enzyme; a nonsynonymous mutation in codon 20; and specified mutations in any of 50 other codons of the glucose-6-phosphatase gene. Another click of the mouse brings up an electronic link to the original scientific paper(s) describing each mutation.

MMBID, *OMIM*, and *HGMD* focus on genes that harbor well-characterized mutations with relatively clear phenotypic effects (salient negative impacts on health, typically). Many additional human genetic disorders have complex etiologies involving multiple genes or gene-environment interactions. These multifactorial diseases include—to mention just a few—diabetes, some mental illnesses, rheumatoid arthritis, and many cancers and cardiovascular diseases. A research challenge is to unveil the mechanistic details of how interacting factors (including hereditary dispositions as well as environmental factors) contribute to each such complex medical disorder.[19]

One recent example of how whole-genome sequencing can help to illuminate the genetic contributions to particular human diseases involved a detailed analysis of a patient with acute myeloid leukemia (a form of cancer).[20] By comparing the complete DNA sequences of normal versus tumor cells in the patient, 10 genes were found to carry new mutations in the tumor cells. Two of these mutations had been described previously and already were known to contribute to tumor progression, but the other eight were previously unsuspected as cancer-associated agents. This study illustrates how personalized genomic analysis may soon supplement more traditional genetic approaches in the diagnosis and perhaps the treatment of patients with complex genetic disorders.

In the modern genomics era, another promising approach to the identification of disease-related genes involves genome-wide association studies (GWASs). The basic idea is to screen large numbers (even millions) of molecular-genetic polymorphisms in each of a large number (thousands) of people and examine the data for statistically significant associations between particular genetic variants and particular human genetic disorders. Toward that end, an International HapMap project was launched in 2002, the goal being to identify genetic variants across the human genome and search for statistical associations that might help to pinpoint the genetic alterations contributing to various complex diseases having at least a partial genetic basis. Since that time, scores of publications from HapMap and related scientific projects have described genetic variants associated with a wide array of clinical conditions and genetic diseases including asthma, atrial fibrillation, breast cancer, celiac disease, colorectal cancer, coronary disease, Crohn's disease, gallstones, glaucoma, inflammatory bowel disease, lupus erythematosus, macular degeneration, multiple sclerosis, obesity, prostate cancer, restless leg syndrome, rheumatoid arthritis, and type 1 and type 2 diabetes. Using this general approach, by early 2008 scientists already had identified more than 150 relationships between common genetic variants and various multifactorial human diseases.[21]

Humanity's Genetic Burden

In his path-breaking treatise one century ago, Archibald Garrod described four inborn errors of human metabolism. Today, thousands of human metabolic abnormalities—often with well-defined clinical symptoms—are known, and the list is growing rapidly as genomic research accelerates. Table 2.2 provides thumbnail descriptions of just a few representative genetic diseases associated with well-characterized metabolic disorders. Overall, deleterious mutations have been documented in each of about 3,000 of the more than 20,000 human genes currently recognized,[22] and as genomic research expands, the list of

TABLE 2.2 Brief synopses illustrating the diversity of genetic errors of human metabolism

Alzheimer disease. A common form of progressive dementia of the elderly; associated with lesions in any of several genes, including mutations in a locus on chromosome 21 that lead to accumulations of amyloid (starch-like) plaques in the brain; affects about four million people in the United States alone.

hypophosphatasia. Symptoms include deformed bones and premature loss of deciduous teeth in children; due to recessive mutations (in a gene on chromosome 1) that compromise the body's ability to mineralize bones; occurs in low frequency worldwide, but is notably prevalent in Mennonite families in Manitoba, Canada.

maple syrup urine disease (a type of aciduria). Symptoms in severe cases result in neonatal brain disease and death, but milder forms can be treated by dietary restrictions on the intake of amino acids that the patient cannot metabolize; named for the maple syrup odor of patient's urine; incidence about one in 185,000 infants worldwide but much higher in some populations; often caused by mutations in a gene on chromosome 19.

Marfan syndrome. Symptoms include instability in the lens of the eye, pulmonary difficulties, susceptibility to hernias, curvature of the spine, spiderlike fingers, and disproportionately long limbs; due to mutations in a fibrillin gene on chromosome 15; incidence is about one person in 10,000.

Menkes' disease. Symptoms include mental retardation, seizures, loose skin, low body temperature, and early childhood death; like Wilson's disease (see below), caused by mutations in a copper-processing gene (on the X chromosome); symptoms arise from an insufficiency of copper in the patient's body.

retinitis pigmentosa. A suite of diseases characterized by degeneration of the eye's retina, first indicated as an inability to see well in poor light and often eventuating in blindness by mid-life; can be caused by mutations in any of several genes, including particular loci on chromosomes 3, 6, 7, 8, 11, 14, 16, and the X.

porphyria. Any of numerous disorders related to a person's reduced capacity to produce hemoglobin (the oxygen-carrying molecule in blood); symptoms range from mild to severe and often include anemia, insomnia, altered consciousness, and intractable pain; can result from mutations in any of several loci involved in the pathways of hemoglobin metabolism.

Salla disease. Symptoms include poor muscle tone and uncoordinated body movements starting at about six months of age, loss of capacity to produce (but not to comprehend) words, mental retardation, shortened life span; due to mutations (in a gene on chromosome 6) that disable a person's ability to process and store sialic acid; the disease is concentrated in northeastern Finland.

Tangier Island disease. Symptoms include orange tonsils, a yellowing of the cornea with age, and peripheral neuropathy that can prevent patients from feeling pain in their arms; occurs worldwide in low frequency, but dramatically higher incidences at a few locations; due to mutations in a gene on chromosome 9 that codes for a protein that plays an important role in transporting high density cholesterol across cell membranes.

Wilson's disease. Symptoms include mania, tremors, joint inflammation, kidney damage, and weakening of the heart; caused by any of more than 200 known mutations in a copper-transporting ATP-ase (ATP7B) gene located on chromosome 13; the symptoms result from too much copper (rather than too little copper, as in Menkes' disease; see above) in a patient's body.

xeroderma pigmentosum. A clinical hallmark is the early onset, typically in childhood, of skin cancers; can be caused by mutations in any of several genes that otherwise contribute to a cell's ability to repair damages from ultraviolet light.

---

disease-causing mutations in protein-coding genes is almost guaranteed to increase substantially.

Genetic disabilities range from the mild to the deadly. One of the most horrific is Lesch-Nyhan syndrome, caused by point mutations in the gene that encodes the enzyme hypoxanthine-guanine phosphoribosyltransferase. Affected children have neurological dysfunctions that lead to compulsions for vomiting and self-mutilation (such as chewing away their lips and fingers, or stabbing their faces and eyes with sharp objects). The afflicted are mentally retarded but have understanding eyes, feel the pain, and are conscious of their uncontrollable behavior.

Lesch-Nyhan syndrome hardly seems like the kind of outcome that would be countenanced by a loving all-powerful Diety. Nor do any other genetic disorders, including some that are truly bizarre. For example, trimethylaminuria (or TMAU) is an incurable metabolic abnormality that can make a person stink like rotting fish.[23] As a result of particular mutations in a gene for a liver enzyme that otherwise breaks down trimethylamine (a standard but smelly by-product of digestion), the afflicted emit a body odor that is repulsive. TMAU has no serious health effects per se, except that patients often find themselves understandably isolated, lonely, and depressed. Researchers estimate that worldwide about 0.1–1.0% of the human population (i.e., millions of people) have mild to serious cases of TMAU.

"Genetic load" is the phrase used to describe the collective burden of genetic disorders in the human population. This genetic millstone is heavy.[24] Gross genetic defects (including deleterious chromosomal mutations as well as metabolic errors per se) are the major known cause of miscarriages in pregnant women from developed countries. About 5% of newborns have a recognizable birth defect, and in at least half of those cases deleterious genes are primarily responsible. Among post-natal infants in developed nations, approximately 30% of mortality is due to overt genetic disorders. About 30% of pediatric admissions to hospitals have a predominant genetic etiology, as do 10% (conservatively estimated) of adult admissions. Apart from routine aging (which itself is genetically dictated to a considerable degree), at least 10% of adults are additionally debilitated by one or more hereditary ailments.

Impressive though such figures are, they grossly underestimate humanity's total genetic load for several reasons. The tallies do not include predilections for diseases such as various cancers, circulatory disorders, obesity, and other multifactorial disorders for which hereditary components are often known or suspected but may remain ill defined mechanistically. They neglect genetic defects that lower fertility. They also disregard severe genetic defects that display so early in human development that the resulting deaths of fertilized eggs or nascent embryos in the womb are not properly recognized as miscarriages.

Finally, standard tallies of genetic load fail to include genetic defects that arise in somatic cells (as opposed to germ-line cells that produce eggs or sperm) and are not transmitted across the generations. All of the mutations described above arose in germ-line cells (those that produce eggs or sperm) of prior human generations and were inherited by the unlucky descendants. But some mutations that arise in somatic cells also have horrendous health consequences for the individual. Cancers provide classic examples.[25] Every cancer begins when a single somatic cell crosses a threshold of genetic alterations such that its capacity for properly regulated cell division is lost. Although a person may have inherited some cancer-predisposing mutations from his or her parents, the additional genetic changes that cause a particular somatic cell to cross the critical cancer threshold typically arise during an individual's lifetime. Indeed, like hitting a negative jackpot, every person will sooner or later contract a cancer if he or she lives long enough.

ERRORS AND FORGIVENESS

A proverbial sentiment is that "To err is human, to forgive is divine." If the kinds of harmful mutations described above are to be attributed to an intelligent and otherwise revered agent (i.e., an omnipotent deity), then the popular phrase needs revision: "to err is divine, to forgive is human." Few people would blame a loving and all-powerful God for purposefully inventing deleterious mutations; that would be blasphemous.

Mutational Mistakes and Corrections

The general scientific explanation is that deleterious mutations are inevitable by-products of natural molecular genetic operations. Each cellular division in the human body necessitates that three-billion-plus nucleotide pairs in the genome replicate and distribute properly to two daughter cells. The fact that the resultant DNA molecules are not always perfect copies of the originals merely reflects the extraordinary difficulty of the replicative operations, coupled with multitudinous opportunities for mistakes (across billions of nucleotides within each of billions of cellular divisions during each human lifetime). During the intricate process of genome replication—in somatic cells or those of the germ line—one or a few nucleotides occasionally are misincorporated. This is not a hypothetical argument; de novo harmful mutations have been documented routinely in all species that have been carefully examined, including humans.

That these are bona-fide molecular "errors" is also not under scientific dispute. Perhaps the most objective evidence for this conclusion is provided by the cellular machinery itself, which seems to go to great lengths to screen and repair DNA mutational damages.[26] Inside the cells of various organisms are legions of enzymes and multiple biochemical pathways that carry out the following kinds of molecular repair: photoreactivation (for fixing pyrimidine dimers that often arise from exposure of nucleic acids to ultraviolet light); nucleotide excision repair (which also works on base dimers); recombinational repair (which repairs DNA via strand exchanges from nonmutated daughter chromosomes); base-excision repair (which identifies and excises aberrant bases); mismatch repair (which recognizes and replaces improperly matched bases with correct partners); inducible repair (which screens for bases that have been chemically modified, for example, by methylation); and SOS repair (which swings into action to renovate single-stranded gaps or the presence of DNA degradation products). Like quality-control inspectors in a factory assembly line, these cellular mechanisms are diligent, but not infallible. The fact that cells employ such extensive and conscientious workforces is itself testimony to the conclusion that mutations create genuine biological defects that are not merely some negative prejudice in the eye of the human beholder. The conventional scientific explanation is that genomes have evolved the capacity to screen and correct mutational errors, but the mechanisms do not always perform perfectly. An alternative but nonscientific explanation, I suppose, is that an intelligent designer has fabricated post-hoc molecular systems, which themselves are imperfect, to partially compensate for molecular flaws inherent in his original biological blueprints.

Population Genetic Excuses

The field of population genetics can predict the frequency of a particular human genetic disorder when the selection coefficient (s, the intensity of natural selection against the defective allele) and the mutation rate (μ) to that allele are known or suspected. For a recessive genetic disorder, the expected equilibrium frequency of the deleterious allele is $q = (\mu/s)^{1/2}$, and for a dominant disease the expected equilibrium allele frequency is $q = \mu/s$. Thus, if a mutation is lethal ($s = 1$) when homozygous and recurrently arises at rate $\mu = 10^{-6}$, its predicted population frequency is $q = (\mu/s)^{1/2} = 0.001$; and if the mutation is lethal even when heterozygous, its population frequency is simply identical to its origination rate ($q = \mu/s = 0.000001$). Many horrible human genetic disabilities with single-gene etiology—such as cystic fibrosis—are sufficiently rare in human populations that no scientific explanation beyond recurrent mutation may be required. Some representative examples are listed in table 2.3.

Other single-gene disorders are far more common and thus demand further explanation. One well-understood example involves sickle cell disease, which threatens the lives of individuals who carry two copies of a sickle-cell allele (S) at a hemoglobin gene. The S allele reaches frequencies of 20% or higher in some African populations, meaning that at least 4% of the populace suffers from this painful and dangerous disability. How did the S

TABLE 2.3 Frequencies in human populations of several representative single-gene disorders

autosomal dominant disorders	
achondroplasia	1 in 50,000
acute intermittent porphyria	1 in 15,000
hereditary spherocytosis	1 in 5,000
Marfan syndrome	1 in 20,000
myotonic dystrophy	1 in 10,000
osteogenesis imperfecta	1 in 20,000
tuberous sclerosis	1 in 15,000
von Willebrand disease	1 in 8,000
autosomal recessive disorders	
albinism	1 in 10,000
Friedreich ataxia	1 in 75,000
phenylketonuria	1 in 12,000
spinal muscular atrophy	1 in 10,000
Wilson disease	1 in 50,000
X-linked disorders	
hemophilia	1 in 10,000 males
hypophosphatemic rickets	1 in 20,000
testicular feminization	1 in 64,000

Source: From Beaudet, A. L., C. R. Scriver, W. S. Sly, and D. Valle, 2001, Genetics, biochemistry, and molecular bases of variant human phenotypes, in *The Metabolic and Molecular Bases of Inherited Disease,* Shriver, C. R. and six others (eds.), McGraw-Hill, New York.

allele become so common? The answer lies mostly in a health benefit—resistance to malaria—that *S* confers when heterozygous. In malaria-infested regions, heterozygotes have a fitness advantage over normal (*A/A*) homozygotes, and they also have an advantage over *S/S* homozygotes by virtue of near freedom from sickle cell disease. In such "heterotic" or "overdominant" situations, where heterozygotes have higher average genetic fitness than either class of homozygote, natural selection can act to stabilize a deleterious allele at far higher population frequencies than would be achieved by recurrent mutation alone.[27] More precisely, the theoretical equilibrium frequency of the *S* allele is predicted to be $q = s / (s + t)$, where s and t are selection intensities against *A*/*A* and *S*/*S* homozygotes, respectively. For example, if *S*/*S* homozygotes are severely debilitated ($t = 0.8$) and *A*/*A* homozygotes show a 20% fitness reduction ($s = 0.2$) relative to heterozygotes, then the expected equilibrium frequency of *S* is $q = 0.20$ (a value that is quite close to empirical reality).

Sickle cell disease illustrates how a gene that confers a fitness advantage to some people can simultaneously relegate other individuals in the same population to a lifetime of sickness and misery. Is this "fair" by any moral or theological standard? Seemingly no. At each conception (long before fetal development and birth), a mere toss of the Mendelian genetic dice determines who will thrive in malarial regions (heterozygotes for the *S* allele) and who will suffer gravely (*S*/*S* homozygotes) from sickle cell disease. We all know that God does not play dice with the universe,[28] so theologians and others may wish to thank natural evolutionary processes for stepping forward to take the blame.

In some cases, deleterious alleles may drift to high frequency in human populations with little or no impetus from natural selection. This can occur, for example, when a population is founded by a small number of individuals some of whom happen to carry a few otherwise rare mutations. Such "founder effects" are on display in the Ashkenazi Jews, who have unusually high frequencies of recessive alleles for several genetic disorders including Tay-Sachs disease and Gaucher disease (table 2.1). Ashkenazi Jews apparently experienced a "bottleneck" (a severe reduction in population size) at A.D. 75 (at the beginning of the Jewish Diaspora) and again between A.D. 1100 and A.D. 1400.[29] Centuries later, many descendants continue to pay a heavy genetic price for the deleterious alleles that their founders happened to carry. Another example of this sort involves Salla disease (table 2.2), which is disproportionately concentrated in the peoples of northeastern Finland. Finland was colonized only 2,000 years ago, and as recently as the late 1600s the population went through a severe decline, thus likewise permitting the genetic drift to high frequency of some mutations that natural selection might otherwise have kept in check.

Some genetic disorders become visible only in particular settings that create a critical gene-environment mismatch. A classic case in point involves the incapacity of the human body to produce vitamin C. This is of no health consequence when ascorbic acid is available from fresh fruits and vegetables, but it becomes a serious genetic disability (leading to scurvy and death) when access to vitamin C is limited, as was often true centuries ago for sailors on long voyages. Another example is provided by X-linked hemophilia, which as mentioned earlier was prevalent in the descendants of Queen Victoria. Hemophilia is not necessarily lethal, but it can quickly become so if accidents or other illnesses initiate uncontrollable bleeding in an afflicted individual (in the royal family, Prince Alfonso, Prince Gonzalo, and Viscount Trematon all died following automobile accidents).

Many genetic disorders occur in high frequency in human populations because their deleterious effects typically are delayed beyond reproductive age. Among countless examples are the cardiovascular diseases, cancers, and many late-onset neurological conditions such as Alzheimer disease, all of which have documented genetic components. Indeed, senescence and nonaccidental death can themselves be interpreted as genetic disabilities, essentially universal to life. Whatever agent sponsors these phenomena must be blind or uncaring about the elderly. Natural selection certainly fits the profile because genetic defects that do not express until late in life usually have little or no negative impact on a person's reproductive success (the ultimate currency of genetic fitness). Does an omnipotent intelligent designer also fit the agent's profile? The question must be asked.

Other Explanations

The mutational process itself is prerequisite for the continuance of life. Without new mutations (some of which happen to be beneficial in particular environments), evolution eventually would cease. But natural selection has no long-term vision, no known capacity to anticipate or plan for the future. Instead, it is a myopic, opportunistic force of nature that operates generation by generation on the different genetic fitnesses of individuals (and the genes they house). Indeed, natural selection can and often does paint its biological products into evolutionary corners from which there is no escape. This routinely happens, for example, when a population evolves a special adaptation to a particular ecological setting that later disappears. Population extinction may then ensue. Yet life on Earth, writ large, has gone on continuously across four billion years. Is this mere good fortune, or does some broader directive agent somehow foresee the future and arrange for life's endurance? With regard to mutations, is the resulting genetic variation itself a part of some foresighted cosmic plan that ensures never-ending adaptive flexibility in the face of inevitable environmental challenges?

Actually, the mutational process itself may be one of the most promising (but ironically neglected) arenas for possible evidence of Intelligent Design. Evolutionary biologists have little more than default explanations for mutations, which are thought to arise, mechanistically, simply as unavoidable mistakes despite a cell's best efforts (normally rewarded by natural selection) to prevent and repair them. Granted, any species that evolves an absolutely perfect DNA repair mechanism probably would go extinct someday as its genetic variability gradually dwindles, but this would be a long-term eventuality that natural selection could not anticipate. Also granted, natural selection can and does operate occasionally at the group level (in this case via the differential extinction of species), so perhaps evolutionary extinctions help to account for why no species alive today is mutation free. Nevertheless, in general, evolutionary explanations for observed mutation rates (which seem well suited to long-term evolution despite high immediate fitness costs) appear to have some aura of intellectual arm-waving that the ID movement might perhaps exploit.

On the other hand, mutations may not be the ideal topic for proponents of Intelligent Design to tout. As this chapter has demonstrated, most de novo mutations range from neutral to highly deleterious for human health, and collectively they leave in their wake countless shattered bodies and destroyed lives (including those of untold numbers of early human embryos). These are probably not the kinds of biological outcomes that one would wish to attribute to the direct hand of an all-powerful and loving God.

CHAPTER 3

Baroque Design: Gratuitous Genomic Complexity

The metabolic disorders discussed in chapter 2 result from deleterious mutations that alter an otherwise standard gene product required by the human body. Perhaps life's architects (whether natural or supernatural) can be excused for these occasional molecular disasters if we assume that harmful mutations are unavoidable aberrations or glitches in a human genome that otherwise was designed and constructed to near perfection. This rationalization would lose its force, however, if the standard human genome were found to contain pervasive architectural and engineering flaws in addition to the many unfortunate mutational departures from the prototype.

For natural theologians in centuries past, as well as to strict adherents of the evangelical movement known as Creation Science in the modern era, biotic complexity is *the* hallmark—*the* unquestionable signature—of intelligent design. However, gratuitous or unnecessary biological complexity—as opposed to economy of design—would seem to be the antithesis of thoughtful organic engineering. A central theme of this chapter (and chapter 4) is that genomic complexity at its most basic molecular level is needlessly baroque, with oft-disastrous consequences for human health.

This was not a foregone scientific conclusion. Before powerful laboratory technologies, including rapid DNA sequencing, became widely available in recent years, most molecular features of the human genome were hidden from direct empirical appraisal, and a general presumption among biologists was that genomic structures and operations would prove to be streamlined and efficient. But the molecular data are now in, and the scientific pictures they paint are both surprising and clear: the human genome is a Byzantine contrivance that departs dramatically from what would seem to be optimal design. Numerous features of the human genome, like those of other multicellular organisms, thus give every indication of having been shaped not directly, ex nihilo, by an intelligent agent, but rather were accumulated stepwise by sloppy tinkering forces.

SPLIT GENES

Geneticists had long supposed that each protein-specifying gene in humans and other animals was a straightforward linear sequence of nucleotides coding for a particular polypeptide. But in 1977, research teams led by Richard Roberts and Phillip Sharp independently announced a startling discovery: protein-coding genes routinely include noncoding sequences interspersed with the coding segments. The intragenic spacers between coding sequences became known as introns and the coding regions henceforth were termed exons. Thus, each standard gene in a eukaryotic organism (any creature whose cells contain nuclei) consists of alternating exons and introns. This unanticipated discovery of "split genes" earned Roberts and Sharp a Nobel Prize in 1993.

INTRONS

The human genome (like those of most other eukaryotic species) is rife with introns, housing more than 150,000 of them in total. Although some loci have more than 20 introns, a typical protein-coding gene comprises about eight exons averaging approximately 160 base pairs each, interspersed with a comparable number of introns of mean length 4,800 base pairs. The DNA sequence in a standard protein-coding gene is thus about 3% exons and 97% introns. In other words, exons are like scattered sentences of informative genetic code embedded in 30-fold larger paragraphs of mostly noninformational twaddle.

Somatic cells themselves seem to view introns as unnecessary script, as judged by the effort they expend in recognizing and discarding intronic sequences during the multistep process of polypeptide production from a functional gene. First, in the cell nucleus, exons and introns of a gene are jointly transcribed into pre-messenger (pre-m) RNA molecules. The cell then utilizes large ribonucleoproteins—spliceosomes—to biochemically remove all intron-derived segments from each pre-mRNA and subsequently splice each gene's exons end-to-end to generate mature mRNAs (figure 3.1). Each spliceosome is a complicated structure of proteins complexed with suites of small nuclear (sn) RNAs. Remarkably, approximately 1% of all known genes in the human genome encode molecular products that our cells employ to build the spliceosomes and conduct splicing operations on pre-mRNA. After the spliceosomes have finished their work, mature mRNAs are exported to the cell cytoplasm where they will serve as templates for protein translation on ribosomes.

All of this cellular rigmarole for pre-mRNA processing is necessitated by the presence of introns. A naïve observer might therefore conclude that introns are an obligatory element of genomic design, probably for some inherent architectural reason relating to proper cellular function or molecular operation. But in fact the genomes of some prokaryotic organisms (such as certain kinds of bacteria) function and replicate just fine without either the kinds or the vast numbers of introns that eukaryotes carry. Even in eukaryotic organisms, some protein-coding genes lack introns also. This is true not only for all genes in human mitochondrial DNA (see below) but also for some protein-coding genes inside our nuclear genome. Thus, introns are not ineluctable or mandatory features of gene design, even within *Homo sapiens*.[1]

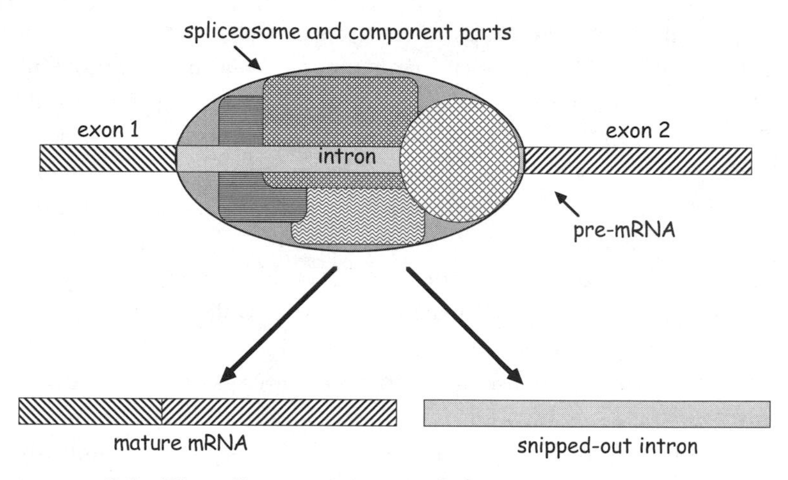

FIGURE 3.1. The spliceosome (not drawn to scale) and its role in snipping out introns from the pre-mRNA molecules that have been translated from protein-coding genes.

There are several reasons to think that cells would be better off without introns, in an ideal world. Introns impose energetic burdens on cells. Introns on average are 30-fold longer than exons and they are transcribed into pre-mRNAs before being snipped out, so they probably extend the time required to produce each mature mRNA molecule by at least 30-fold (compared to the expectation for non-split genes). Even if time is not of the essence to somatic cells, the metabolic costs of maintaining and replicating all the extra nucleotides in introns must be considerable. To these cellular costs must also be added the metabolic expense of generating spliceosomes and running the extensive pre-mRNA processing machinery.

Nevertheless, for the sake of argument let us assume that the metabolic costs imposed by introns are negligible. Do introns otherwise provide evidence of optimal genomic design? No, because pre-mRNA processing also has opened vast opportunities for cellular mishaps in protein production. Such mishaps are not merely hypothetical. An astonishing discovery is that a large fraction (perhaps one-third) of all known human genetic disorders is attributable in at least some clinical cases to mutational blunders in how pre-mRNA molecules are processed.[2] For example, mutations at particular intron-exon borders often disrupt pre-mRNA splicing in ways that alter gene products and thereby lead to countless genetic disabilities including various cancers and other metabolic defects.[3] There is also good evidence from the human genome that the number of introns in a gene is positively correlated with a gene's probability of being a disease-causing agent.[4]

Table 3.1 describes some of the many human genetic afflictions that have been documented (in particular clinical instances) to molecular errors in mRNA splicing at specifiable genes. These range from a variety of neurodegenerative diseases to debilitations of the circulatory, excretory, and other body systems. Many of these devastating genetic disorders begin in infancy or early childhood; others may be deferred to the elderly. To illustrate the horror of some such diseases, consider amyotrophic lateral sclerosis, also known as Lou Gehrig disease (named after the famous baseball player who died of the disorder). This disease affects a part of the nervous system that controls voluntary movements, with the result that muscles get progressively weaker until the victim eventually becomes paralyzed and dies. Patients usually first lose an ability to move their arms, legs, and body. Eventually, muscles in the diaphragm and chest fail as well, leading to slow asphyxiation. The agony is normally prolonged across three to five years, but can last a decade or more before the inevitable death ensues. No cure has been found.

The disorders listed in table 3.1 are a tip of the iceberg, but even a full compilation of such gene-specific splicing disorders would seriously underestimate the total burden that introns impose on human health. Some mRNA processing mutations—in splicesome construction or operation, for example—have pervasive impacts on mRNA splicing by simultaneously disrupting protein production from many genes. Gross mutations of this sort, when expressed in utero, normally are lethal, so the affected embryos or fetuses seldom would survive long enough to be recognized for inclusion in the tally of genetic casualties of RNA processing.

TABLE 3.1 Examples of oft-devastating genetic disorders sometimes caused by mutations that affect RNA splicing (ultimately necessitated by the presence of introns) in human cells

amyotrophic lateral sclerosis. A neurodegenerative disease with symptoms of muscle wasting, weakness, and spasticity that prove ultimately fatal; some cases are due to aberrant processing of RNA molecules from *EEAT* genes that encode particular neurotransmitter proteins.

Ehlers-Danlos syndrome, type IV. An autosomal dominant disorder often characterized by joint hypermobility, thin and translucent skin, and a proneness to spontaneous rupture of the bowel and large arteries; some cases have been tied to mutations at splice-site junctions in a collagen-encoding gene.

elliptocytosis. A disorder in which patient's red blood cells are elliptical rather than biconcave, with symptoms in various cases ranging from nonclinical to severe; some instances are due to "skipping" errors at splice junctions in an erythrocytic gene known as SPTB.

frontotemporal dementia with parkinsonism-17 (FTDP-17). A progressive dementia, typically beginning at 40 to 60 years of age, with clinical features including behavioral and cognitive changes, language disabilities, and motor dysfunctions; due to splicing errors among exons of the *tau* gene, on chromosome 17, that produces a microtubule protein involved in vesicular transport in neurons.

ornithine transcarbamylase deficiency. A deficiency in the body's capacity to excrete ammonia (a breakdown product of proteins), often beginning in early infancy and in severe cases resulting in coma, brain damage, and death; some cases result from mutations at splice junctions in an OTC gene that encodes ornithine transcarbamylase.

spinal muscular atrophy. A progressive degeneration of spinal cord neurons in children, and among the most common of all genetic causes of childhood mortality; some cases are due to a silent nucleotide transition—in a motor neuron gene on chromosome 5—that leads to inappropriate splicing of exons from this locus.

Wilm's tumor. A rare kidney cancer or nephroblastoma that mostly affects children; some cases are due to aberrant forms of a tumor suppressor protein that arise via inappropriate alternative splicing of exons from a WT gene; this is just one of many cancers for which some particular cases are known to involve similar kinds of RNA splicing anomalies.

others. Among the many other genetic disorders stemming in some cases from RNA-splicing errors are various haemophilias and thalassaemias, acatylasaemia, adrenal hyperplasia, analbumiemia, cystic fibrosis, Lesch-Nyhan syndrome, LPL deficiency, osteodystrophy, phenylketonuria, porphyria, protein C deficiency, retinoblastoma, and Tay-Sachs disease.

Evolutionary Enigmas

Ever since their discovery, split genes and introns have been a scientific enigma, perplexing evolutionary biologists and medical geneticists alike. One hotly contested issue concerns the evolutionary dates of intron origin. One scientific camp holds that introns were ancestral genomic features that originated billions of years ago in primordial prokaryotes—only to be lost later in most prokaryotic descendants. A standard corollary of this view is that introns also played a major role in early protein evolution via exon shuffling (as described below). A competing hypothesis holds that introns were genetic afterthoughts that populated eukaryotic lineages much later in Earth history and thus played no role in protein evolution at life's outset. This scientific debate continues today.[5]

Challenging questions also exist about the molecular mechanics of intron origin.[6] One hypothesis is that introns are ghosts of gene fragments past. Such ghosts may have accumulated through the mutational silencing of redundant DNA sequences that originated as tandem duplications of portions of protein-coding genes. Such tandem duplication events are common in eukaryotic genomes (chapter 4), but the unanswered questions are whether and how often introns mechanistically originated by this route. A second hypothesis addresses intron proliferation more than origins per se. It posits that each new intron is born when a preexisting intron is released from pre-mRNA and then reintegrates elsewhere in the genome. A third hypothesis is that introns are evolutionarily related to mobile elements or "jumping genes."

A mobile element is a quasi-autonomous piece of virus-like DNA that can proliferate in the genome of a host lineage by replicating and hopping from one chromosome site to another (see chapter 4). One suggestion is that mobile elements were the evolutionary predecessors of introns.[7] A different interpretation is that introns seldom arose from transposable elements and instead that mobile elements evolved when their ancestors learned to take advantage of preexisting splicing mechanisms for introns.[8]

To date, no clear scientific consensus has emerged regarding the evolutionary origins of introns. Instead, each hypothesis mentioned above has various lines of empirical evidence in its favor, and against. Furthermore, not all of the hypotheses are mutually exclusive. It is probably the case that introns arose multiple times in evolution and by more than one molecular route.

Regardless of their dates and mechanisms of origin, introns are highly consequential for eukaryotic cells. One functional benefit from split genes is immediate and direct. After intron sequences have been removed from pre-mRNA, the remaining coding sequences from different exons are sometimes spliced together in alternative formats, resulting in multiple proteins with varied structures and functions. This process, known as alternative splicing (figure 3.2), means that N genes can yield $>N$ different proteins. In the human genome, more than 75% of the protein-coding genes express alternatively spliced variants.[9] If not for alternative splicing, the genomes of humans and other eukaryotic species would need to house many more protein-coding genes than they actually do.

An exon sometimes encodes a discrete functional domain of a protein, such as an active site or other recognizable subunit of an enzyme. This observation led to the provocative exon-shuffling theory that might help to explain the evolutionary origins of new genes themselves.[10] The basic idea is that during the course of evolution, genetic recombination among prefabricated genetic modules (exons) occasionally results in the fortuitous assembly of mosaic genes with functional capabilities not possessed by their predecessors (figure 3.2). The modular design of protein-coding

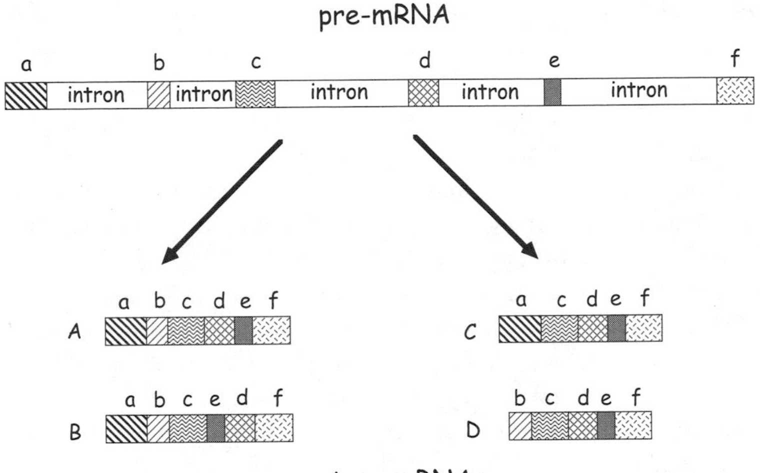

FIGURE 3.2. Alternative splicing. Outcome A faithfully reflects the order of exons in the gene, outcome B is an example of "exon swapping" in which exons d and e were inverted, and outcomes C and D are examples of "exon skipping" in which exons b and a, respectively, were excluded during the production of mature mRNA. Alternative splicing involves RNA and is a developmental phenomenon. Exon shuffling is quite analogous but involves DNA and is an evolutionary phenomenon (see text). In the case of exon shuffling, recombination events reorder exons in the genes themselves.

genes, in which exons and introns are interspersed, should greatly facilitate this domain-shuffling process and thereby lead occasionally to functionally novel proteins. Even if efficacious exon-shuffling events are individually rare, large genomes and vast time scales make the evolution of novel genes and gene functions almost inevitable by this genetic mechanism.

Theological Enigmas

Earlier I discussed some of the many operational costs that split genes precipitate, ranging from the precious time and energy that each cell must expend just to maintain introns and run the complex RNA splicing machinery, to the many instances in which RNA processing goes seriously awry and thereby produces countless metabolic disorders and premature deaths. These functional shortcomings beg the question: Do introns debunk intelligent design? Or, alternatively, do the functional benefits of introns compensate for their evident costs in ways that might point to intelligent and caring genomic craftsmanship, as opposed to tinkered evolution by natural forces?

Alternative splicing allows a cell to produce a plethora of different proteins from relatively few genes, and such structural and functional efficiency might suggest conscious thought behind genomic design. But any general exigency for streamlining in the human genome is otherwise flatly contradicted by a mountain of empirical evidence. As described in chapter 4, noncoding repetitive sequences—"junk DNA"—comprise the vast bulk (at least 50%, and probably much more) of the human genome. Even if we confine our attention to protein-specifying genes, 97% of such DNA sequences consist of introns rather than codons. If an intelligent designer had dispensed with introns, he could have packed about 30-fold more protein-coding genes into the available gene space[11] while simultaneously avoiding all of the operational flaws inherent in the current system of pre-mRNA processing. Or, if introns are for some reason an indispensable part of optimal gene design, surely an omnipotent engineer could at least have made them much smaller so that they would squander less of the cell's precious resources. Thus, a desire for genomic efficiency via alternative splicing cannot be accepted as an intelligent designer's motive for inventing split genes.

Conversely, the complexities of split genes and splicing operations cannot be taken as strong evidence of intelligent design either, because molecular blunders in RNA processing are unquestionably responsible for multitudinous metabolic disorders and vast human suffering. If a higher intelligence directly instigated this error-prone genomic complexity, that agent was either highly fallible as a genetic engineer or largely unconcerned about people's genetic health.

In conclusion, the present-day molecular and operational details of split genes do not inspire confidence that either the complexities or the acknowledged benefits of RNA processing register the immediate and explicit designs of a loving omnipotent intelligence. With respect to introns, perhaps the most viable remaining hypothesis for intelligent design is that the engineer showed foresight in inventing exon shuffling for the long-term good (evolutionary persistence) of life. But even if this notion were true (which seems scientifically implausible), an intelligent designer who merely established suitable conditions for natural biological evolution would be quite unlike the traditional God of the Creation Science movement.

GENE REGULATION AND RNA SURVEILLANCE

MECHANISMS

Each protein-coding gene or "structural gene" also has adjoining (*cis*) regulatory sequences that help to modulate when (during development) and where (in different tissues or organs) the gene is expressed. Most notable is a core promoter, usually several dozen base pairs long, containing nucleotide sequences to which particular suites of proteins known as transcription factors bind, to be joined by RNA polymerase molecules that catalyze the fabrication of pre-mRNA from the adjoining structural gene.[12] Other regulatory sequences called enhancers and silencers, sometimes thousands of nucleotides upstream or downstream from the core promoter, additionally boost or inhibit transcription. Each gene may have several enhancers and silencers; these can be shared among genes, but different genes have different combinations. The enhancers and silencers influence transcription via their connections to large families of activator and repressor proteins that transpond regulatory signals to RNA polymerase via co-activators and other proteins.[13] Distinct batteries of transcription factors and their molecular associates operate in different types of cells, thereby helping to explain how different tissues and organs within an individual can have very different patterns of gene expression despite sharing the same underlying genome. Figure 3.3 is a simplified diagram of some of the regulatory elements of gene transcription.

Once a protein-coding gene has been turned on by appropriate regulatory signals, and mRNA has been transcribed, mechanisms of "RNA surveillance" spring into action.[14] For any of a variety of reasons, some mRNA molecules have become mistakenly truncated or otherwise blemished in ways that would prevent their effective translation into a useful polypeptide. Somatic cells monitor for such defects and actively degrade many of the damaged mRNA copies. This makes good design sense; if faulty mRNAs were not destroyed, dysfunctional rogue proteins might appear in cells far more often they do. With respect to correcting genetic errors, RNA surveillance is the RNA-level analogue of the various DNA repair mechanisms (see chapter 2) that operate directly at the level of genes.

One major RNA surveillance pathway is known as nonsense-mediated mRNA decay, or NMD. It specializes in recognizing

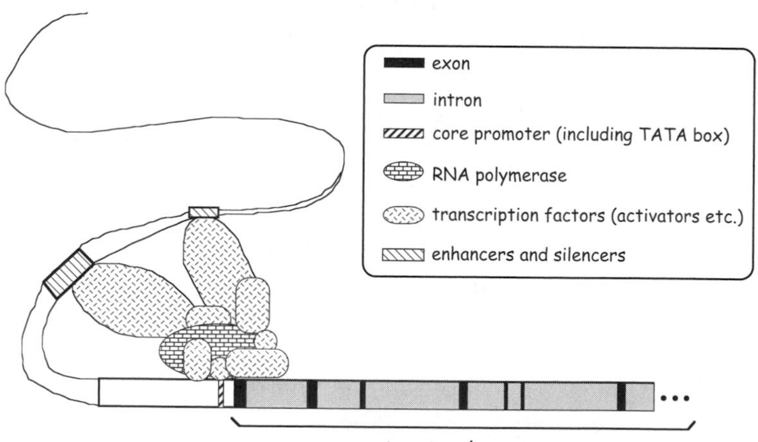

FIGURE 3.3. Schematic (not drawn to scale) of some of the molecular machinery involved in the transcriptional regulation of gene expression.

and degrading mRNAs that are truncated by virtue of nonsense mutations in their respective genes. A nonsense mutation is one that generates a stop codon where it does not belong, thereby in effect truncating the coding sequence prematurely ("chain termination"). The NMD biochemical pathway facilitates the recognition and destruction of such truncated mRNA molecules.

In a broad definitional sense, the genetic regulation of protein-coding genes can also occur at any post-transcriptional stage of protein production, including pre-mRNA editing, the exportation of mature mRNAs from the nucleus, differences in the stability and transport of mRNA molecules after they have reached the cytoplasm, factors impinging on the translation process by which polypeptides are constructed from mRNA on ribosomes, polypeptide assembly into functional proteins, and post-translational modifications or degradations of the proteins themselves (figure 3.4). Many of these regulatory mechanisms involve complex biochemical pathways, and collectively they necessitate major expenditures in cellular effort and molecular materials. Probably 50% or more of all coding genes in the human genome could be considered to play some direct or indirect regulatory role in development—for example, in cell signaling and communication, control over gene expression per se, or influences on cell division, structure, or motility.

Protein kinases provide a leading illustration of post-transcriptional regulation in eukaryotic cells. Kinases are enzymes that phosphorylate (add phosphate groups to) and thereby alter the activity of substrate molecules. The human genome contains about 518 functional protein kinase genes (about 2% of all protein-coding loci) that can be arranged into several dozen functional families and subfamilies of loci, all of which arose—under an evolutionary interpretation—from successive gene duplication events (see chapter 4) across the long history of vertebrate animals.[15] By altering the activity profiles of proteins, kinases exert regulatory control over numerous cellular processes including

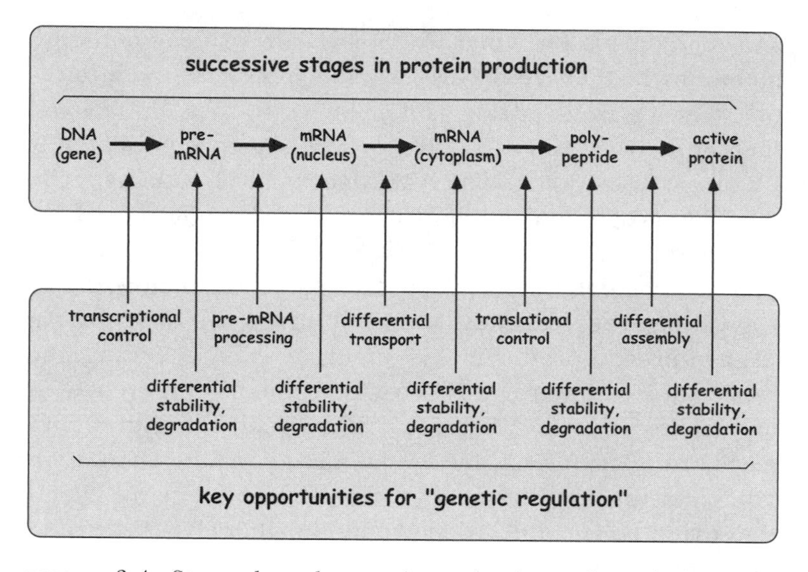

FIGURE 3.4. Stages along the protein-production pathway that provide key opportunities for genetic regulation.

metabolism, cell-cycle progression, cell movement and differentiation, physiological homeostasis, functioning of the nervous and immune systems, and signal transduction (mechanistic pathways by which chemical or other environmental stimuli evoke cellular responses).

Newly discovered micro-RNAs (miRNAs) are another important class of loci involved in post-transcriptional genetic regulation.[16] Each miRNA is a short (about 20-nucleotide) stretch of RNA that can bind to complementary sequences in the messenger RNA molecules of protein-coding genes and thereby inhibit the translation or induce the degradation of specific genetic messages. Although the exact numbers and precise roles of miRNAs in the human genome remain to be fully illuminated, more than 500 such loci already have been identified, and early findings suggest that miRNAs might prove to be major cellular tuners of protein synthesis.

Another interesting class of molecules is long noncoding RNAs, each typically hundreds or thousands of base-pairs long.[17] Recent work on humans and other animals has uncovered thousands of such RNAs from loci that do not encode proteins (or tRNA or rRNA), and such findings beg the question: Why do cells produce these molecules from what otherwise might seem to be junk DNA? Although human cells gainfully employ at least some of these long RNAs (such as "Xist RNA" that helps to regulate gene expression along the full length of the X chromosome), it is currently unclear whether such functional examples are the rule or the exception. Some geneticists posit that long noncoding RNAs will prove to be important regulators of gene expression, whereas others demur on this possibility for now, pointing to countervailing evidence such as the fact that cells seem to destroy long RNAs almost as soon as they are produced. As with any such scientific challenge, it will be epistemologically impossible to prove the null hypothesis that these loci have no functional utility. About all that can be said for now is that more scientific research on long RNAs is needed. Toward that end, the junk-DNA hypothesis can perhaps serve the useful role of "devil's advocate" in stimulating such research.

The various mechanisms described above thus help to orchestrate how particular genes and their protein products are expressed within a cell. Many additional routes to gene regulation also exist, such as how nucleic acid sequences are spatially organized and packaged into chromatin fibers and chromosomes,[18] how DNA molecules are complexed with histone proteins,[19] and the pattern in which cytosine bases in DNA sometimes are modified via chemical methylation. In short, the sheer complexity of structure and function in the genetic regulatory apparatus of cells is not in dispute.

Functional Problems

Regulatory complexity during development is a double-edged sword for any organism. Regulatory mechanisms permit the functional specialization of cells in different tissues and organs, as well as allow potentially appropriate metabolic cellular responses to external or internal environmental stimuli; but the molecular machineries of gene regulation are metabolically costly and often malfunction with disastrous health consequences.

Improprieties in one or another aspect of gene regulation are responsible for a host of human ailments ranging from particular cases of asthma to various immune disorders, circulatory problems, and heart diseases. Many manifestations of cancer have been traced to aberrant methylation patterns in the promoter regions of particular genes.[20] Thalassemias—genetic disabilities that arise from inadequate supplies of oxygen-carrying globins in the blood—are another large class of metabolic diseases related to problems in gene regulation. By definition, thalassemias entail errors in globin production rather than malfunctions in the globin proteins per se. As recognized by Weatherall and colleagues[21] from molecular evidence that already was available a quarter-century ago, these blood disorders result from "a diverse series of *cis* acting lesions of the globin genes which include deletions, insertions, frame shift mutations, and point mutations involving transcription, messenger RNA processing, initiation, termination, poly-A addition[22] and globin chain stability." Similar statements, properly modified to suit the category of protein, could be made about how regulatory malfunctions underlie many other classes of human metabolic disorders (additional representative examples can be found in table 3.1).

Various members of the large family of protein kinases—key metabolic regulators in cells—are also subject to disorder-producing malfunctions. Collectively, more than 160 different kinases have been implicated in cancers by their common association with particular tumor types, and 80 kinases have likewise been associated, at least provisionally, with various other disease conditions in humans.[15] Although still in its infancy, research on microRNAs similarly suggests that occasional misregulation of miRNA molecules likely contributes to the total pool of human metabolic disorders (including, perhaps, DiGeorge syndrome as well as some cancers).[16]

The cell's RNA surveillance systems are likewise complex and subject to occasional errors or otherwise inappropriate actions.

For example, whereas cases of Marfan syndrome (see table 2.2) often result from constitutive loss-of-function (typically nonsense) mutations, the problems can also arise or be exacerbated by failures of the nonsense-mediated mRNA decay surveillance system to protect against the occasional production of truncated fibrillin proteins. Some genetic disorders may even be accentuated by a "properly operating" NMD. This can happen when a stable nonsense mRNA, despite being partially defective, nonetheless could have generated enough protein to meet a cell's functional demands if it had not been degraded first by NMD. For example, some forms of thalassemia (see chapter 2) would be milder if not for such counterproductive NMD behavior.

Evolutionary and Theological Enigmas

Why an intelligent and loving designer would have infused the human genome with so many potential (and often realized) regulatory flaws is open to theological debate. Any such philosophical discussion should probably include not only the issue of whether the designer was fallible (and if so, why?), but also whether the designer might recognize his own engineering fallibility (as perhaps evidenced, for example, by the RNA surveillance mechanisms that seem to have been deployed to catch some of the numerous molecular mistakes).

From an evolutionary perspective, genetic flaws may seem easier to explain. Occasional errors in gene regulation and RNA surveillance are to be expected in any complex contrivance (such as the human genome) that has been engineered over the eons by the endless tinkering of mindless evolutionary forces: mutation, recombination, genetic drift, and natural selection. Thus, by this argument alone, the complexity of genomic architecture would again seem to be a surer signature of tinkered evolution by natural processes than of direct invention by an omnipotent intelligent agent.

Today, scientific journals and textbooks are rich with examples that interpret the molecular minutiae of various genomic features (including mechanisms of gene regulation) in an evolutionary framework. However, my intent in this book is not so much to trumpet the power of evolutionary explanations for genomic features (this is done at great length in the scientific literature) but rather to raise consciousness about some logical problems with biological inferences based on the precept of ex nihilo intelligent design.

GENETIC IMPRINTING

In mammals including humans, some genes routinely are expressed when inherited from one parent but not from the other. In such cases, an allele (a particular copy of a gene) can have very different effects on progeny depending on whether it was transmitted from the mother or from the father. Scientists have documented such genetic "imprinting" at more than 60 human genes to date, but several hundred loci probably experience the phenomenon in our species. Mechanistically, imprinting usually results from the addition of methyl (–CH3) groups to particular nucleotides during the production of sperm or eggs, resulting in maternal-specific or paternal-specific gene inactivation in offspring. Imprinting is disproportionately prevalent for genes expressed in the human placenta (the organ that enables the transfer of resources from a pregnant mother to her developing fetus) and in the human brain.

Mistakes

A gene known as *IGF2*, which encodes an insulin-like growth factor, offers a clear example of how errors in genetic imprinting can have important consequences for human health. Normally, only the paternal copy of *IGF2* is expressed in offspring. If the father's copy of *IGF2* is silenced (through a biochemical mishap in spermatogenesis), the result is a child with Silver-Russell syndrome, a disorder characterized by abnormally low birth weight and growth retardation. Conversely, if the mother's copy of *IGF2* is expressed (through a biochemical mishap in oogenesis), the result is a child with Beckwith-Wiedemann syndrome, a disorder characterized by high birth weight and symptoms of over-growth. Abnormalities in genomic imprinting are likewise known (or in some cases strongly suspected) to underlie, in whole or in part, several other human metabolic disorders (table 3.2). These include four of the most common problems that can arise during human gestation: preeclampsia, miscarriage, fetal growth restriction, and gestational diabetes.

TABLE 3.2 Examples of metabolic disorders for which genomic imprinting is known to be a contributing factor in at least some cases

Angelman syndrome. A disorder with delayed development and neurological problems including jerky movements, sleep disorders, and seizures; associated with a small gene region on chromosome 15, which when improperly imprinted or missing (from a deletion event) in the father results in this syndrome in progeny (see also *Prader-Willi syndrome*).

fetal growth restriction. Any condition in which a fetus is unable to reach its genetically determined potential body size; can have many different etiologies, including genetic imprinting errors.

gestational diabetes. A form of diabetes (a disease in which the pancreas cannot properly produce or utilize insulin) that occurs in otherwise nondiabetic women during pregnancy; affects about 3–6% of all pregnant women.

miscarriage, or spontaneous abortion. A termination of pregnancy within the first 20 weeks of gestation; can have many different causes, but some cases involve imprinting errors.

Prader-Willi syndrome. A genetic disorder involving short stature, poor motor skills, obesity, underdeveloped sex organs, and mental retardation; associated with a small gene region on chromosome 15 (the same genetic region as for *Angelman syndrome*), which when inappropriately imprinted in the mother results in this syndrome in progeny.

(*continued*)

TABLE 3.2 (*continued*)

preeclampsia. A disorder that affects both the mother and fetus, typically in the second or third trimesters, with rapidly progressing symptoms including sudden weight gain, headaches, and changes in vision; affects at least 5–8% of all pregnancies, and can be fatal.

Rett syndrome. A neurological disorder characterized by autistic-like behavior, language impairment, and mental retardation, beginning in early childhood; caused by male germ-line mutations in an X-linked gene (MECP2) that codes for methyl-CpG binding protein.

Turner syndrome. A genetic disorder in girls often associated with failed ovarian development, webbed neck, drooping eyelids, abnormal nails, heart and kidney defects, and other symptoms; caused by a missing or defective X chromosome; some of the phenotypic features depend upon whether the single inherited X is of paternal or maternal origin.

Scientific evidence also suggests that some widespread psychiatric illnesses including autism and schizophrenia are often linked to imbalanced gene imprinting.[23] It has long been known that various mental illnesses tend to run in families, yet they seldom seem to obey the conventional laws of Mendelian inheritance. The new suggestion is that patterns of imprinted-gene expression impact brain development and that deviations in those imprinting patterns can cause metabolic imbalances leading in exceptional cases to clinical mental illness. If so, where a person resides on a continuous spectrum ranging from autism to mental normalcy to psychosis may be partly ensconced in his or her imprinted genes.

Theological and Evolutionary Excuses

Despite an abundance of theological rationales, the theodicy conundrum persists as to why an intelligent creator would have burdened humans with apparent genomic flaws, including the complications and malfunctions of genetic imprinting. Consider, for example, instances of autism resulting from imprinting errors. Autism is a brain disorder, with no known cure, in which the mentally disabled child displays self-focused behavior, a lack of empathy, and a generally impaired capacity for social interactions. In traditional folklore, children with autism were thought to have been "stolen" from their parents by supernatural entities.[24]

Evolutionary biology has provided an entirely different perspective based on the "conflict theory of genetic imprinting." This hypothesis begins with the realization that each individual in a sexually reproducing species receives genes from two different parents, thus setting a stage for potential gene-by-gene conflict over optimal tactics for representation in succeeding generations. In mammals (where females are the pregnant sex), each sire is presumably under selection pressure to transmit genes imbuing offspring with the proclivity to extract as many resources as feasible from the female (e.g., via the placenta) because this selfishness by offspring comes at no personal cost to the male whereas each dam, who pays the personal cost of gestation, is presumably under selection pressure to transmit genes that predispose an embryo to be satisfied with a more equitable allocation of resources between itself and its mother.[25]

To illustrate, consider the aforementioned *IGF2* gene, which encodes an insulin-like growth factor. Normally, only the father's copy is expressed in offspring, each of whom therefore has only one dose of the gene and tends to grow to normal size. In Silver-Russell syndrome, however, the father's copy of *IGF2* is silenced, so the fetus has less growth factor and grows with abnormal slowness. Conversely, in Beckwith-Wiedemann syndrome, the paternal and maternal copies of *IGF2* are both active, leading to a superabundance of growth factor and abnormally fast fetal growth. In general, a genetic tug-of-war across evolutionary time has led to normal imprinting patterns that represent a compromise between what would most benefit genes transmitted by males and what would most benefit genes transmitted by females.

MITOCHONDRIAL DNA

Mitochondria (singular: mitochondrion) are tiny organelles that reside in the cytoplasm of most human cells (except erythrocytes). Their name derives from mitos meaning thread, and chondria meaning granules. They were discovered more than a century ago by the German pathologist Richard Altmann, who thought that they might be free-living creatures within each cell and accordingly referred to them as "elementary organisms."[26] Scientists now know that mitochondria are indeed descended from elementary organisms within us (albeit not in quite the way that Altmann had envisioned). As we will see later in this chapter, mitochondria apparently are the molecular legacy of a close symbiotic association early in the evolutionary history of life.

Mostly as a consequence of their cytoplasmic housing, mitochondria are maternally inherited in humans and other animals. In any zygote or fertilized egg, the cytoplasm and its molecular contents come almost exclusively from a dam's cytoplasm-rich oocyte rather than a sire's cytoplasm-poor sperm cell. Thus, generation after generation, mothers alone transmit mitochondria to their children (except in rare instances of "paternal leakage"). This matrilineal transmission mode, absent the usual complications of paternal input and genetic recombination that impact nuclear genes, greatly simplifies genealogical bookkeeping—that is, in deciphering how different mtDNA sequences in the human population today trace back through time to shared matrilineal ancestors in our past. In particular, anthropologists have used genetic information from mitochondria to detail human matrilineal origins and trace how our ancestors peopled the planet in the last 200 millennia.[27]

Structure

Mitochondria are the only cytoplasmic organelles in humans to house their own DNA (abbreviated mtDNA), a fact not uncovered until 1963.[28] Eighteen years later, scientists determined the entire nucleotide sequence of one representative human mtDNA,[29] a technological achievement that was astonishing at that time but would be standard procedure today. A prototypical mtDNA molecule in humans is 16,569 base pairs long. Each mtDNA is a closed circle of 37 genes, 22 of which encode transfer RNAs, 13 specify polypeptides, and two encode ribosomal RNAs (figure 3.5). Each mtDNA also includes a control region (CR) with an initiation site where replication of the molecule begins. MtDNA replication is asynchronous with cell division, and frequent, such that hundreds or thousands of mtDNA copies normally inhabit the cytoplasm of each somatic or germ-line cell. Despite these large populations of molecules within cells, most individuals are effectively homoplasmic for mtDNA, meaning that most mtDNA molecules within an individual are genetically identical (except for occasional de novo mutations of recent origin).[30]

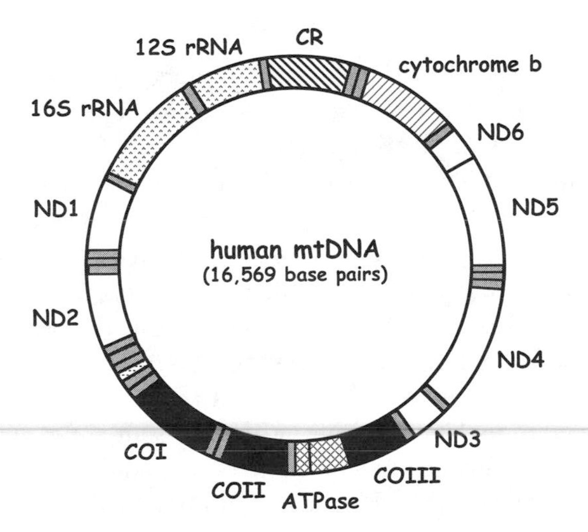

FIGURE 3.5. Schematic representation of a human mtDNA molecule. CR, control region; ND, NADH dehydrogenase; CO, cytochrome oxidase. The 22 stippled regions indicate various tRNA loci.

Unlike most protein-coding nuclear genes, mitochondrial loci have no introns, no noncoding flanking sequences, and no spacer sequences between genes. Mitochondrial genes are thus tightly packed. The entire mtDNA molecule is transcribed as a single unit, and the tRNA sequences that punctuate the protein-coding genes are then cleaved out. These tRNAs subsequently assist in translating mitochondrial mRNAs on special mitochondrial ribosomes—generated from the mitochondrial 16S and 12S subunits—that are distinct from the ribosomes that translate nuclear genes. This distinction is important because the genetic code of mtDNA differs somewhat from the genetic code employed by the nuclear genome (meaning that the nuclear and mitochondrial ribosomes cannot effectively cross-translate).

Table 3.3 summarizes many of the fundamental structural and operational differences between the nuclear and mitochondrial genomes of humans. None of these differences make much sense, except in the light of evolution (see later in the chapter).

Function

Mitochondria are the primary seats of energy production in cells and hence are often referred to as life's biochemical factories or intracellular power plants. As such, the metabolic reactions that occur within and between their walls (mitochondria have an inner and an outer membrane) are responsible for churning out most of a cell's biochemical equivalent of electrical power: ATP (adenosine triphosphate). Mitochondria don't burn fossil fuels; rather, they generate energy by oxidizing hydrogen (derived from the carbohydrates and fats that we ingest). The process entails creating a capacitor (in this case, an electron chemical gradient across the mitochondrial inner membrane) that becomes an energy source for generating ATP and heat. The principal biochemical pathway in mitochondria by which this is carried out is oxidative phosphorylation (OXPHOS), of which the respiratory chain is a key component.

The respiratory chain consists of five enzyme complexes (I–V) plus coenzyme Q and cytochrome c (figure 3.6). Complexes I

TABLE 3.3 Major structural and operational differences between the nuclear and mitochondrial genomes of humans

Feature	Nuclear genome	Mitochondrial genome
size	>3,200,000,000 bp	16,569 bp
no. genome copies per diploid cell	46	thousands
no. genes encoded	~25,000	37
gene density	1 per ~40,000 bp	1 per ~450 bp
introns	plentiful	absent
% coding DNA	~3	~93
codon usage	"universal" code	departures from universal
associated proteins	many histones	no histones
mode of inheritance	generally Mendelian	maternal, asexual
transcription of genes	most done individually	all done collectively
effective genetic recombination	plentiful	rare or absent

Source: Modified from Taylor, R. W and D. M. Turnbull, 2005, Mitochondrial DNA mutations in human disease, *Nature Rev. Genet.* 6, 389–402. The treatment in Taylor and Turnbull was itself modified from Strachan, T. and A. P. Read, 1999, *Human Molecular Genetics*, 2nd edition, John Wiley, New York.

and II oxidize NADH and succinate, respectively; complexes I, III, and IV pump protons to effect an electrochemical gradient; and complex V uses energy from that gradient to synthesize ATP from ADP (adenosine diphosphate). A remarkable fact is that four of these five enzyme complexes are composed of combinations of polypeptides from the mitochondrial and nuclear genomes (figure 3.6). In complex IV, for example, three of the 13 polypeptides are encoded by mitochondrial loci (COI, COII, and COIII) whereas the remaining polypeptides are encoded by nuclear genes. Only in complex II are all of the necessary enzymatic subunits (four in this case) encoded by just one genome (the nuclear).

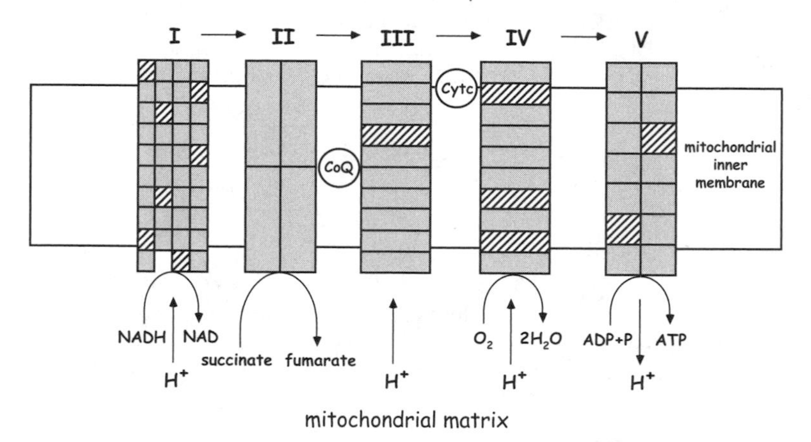

FIGURE 3.6. Simplified diagram of the respiratory chain on the mitochondrial inner membrane (modified from Graff, C., D. A. Clayton, and N.-G. Larsson, 1999, Mitochondrial medicine—recent advances, *J. Internal Med.* 246, 11–23). Shown are the five enzyme complexes (I – V) plus coenzyme Q (*CoQ*) and cytochrome *c* (*Cytc*). For each enzyme complex, each hatched box represents a distinct polypeptide subunit encoded by mtDNA, and each shaded box indicates a polypeptide subunit that was encoded by nuclear DNA and then exported to the mitochondrion (see text).

Nuclear genes are also intimately involved in other basic mitochondrial functions. Indeed, mtDNA does not encode any of the proteins that are directly involved in its own replication, transcription, translation, surveillance, or repair. In short, mtDNA is just a tiny snippet of DNA that by itself would be absolutely helpless, to itself and to the organism in which it is housed. None of this makes any biological sense either, except in the light of evolution (see later in the chapter).

MALFUNCTION

Like every other genetic system that we have considered thus far, the mitochondrial genome is plagued by mutations that often compromise molecular operations. Indeed, on a per-nucleotide basis, mtDNA experiences at least 5–10 more mutations per unit time than do typical protein-coding genes in the cell nucleus.[31] Many of these mutations are of little or no consequence to a person's health, and indeed such selectively neutral mutations provide ideal genetic markers for reconstructing matrilineal relationships within our species. However, other mtDNA mutations have a host of negative effects on a person's health that can range from mildly debilitating to deadly. Clinical disabilities that have been linked to mtDNA mutations disproportionately involve high-energy tissues and organs: brain, the eye and other components of the peripheral nervous system, heart, skeletal muscle, kidney, and the endocrine system.

One of the first human disorders firmly documented to stem from an mtDNA mutation was Leber's hereditary optic neuropathy (LHON), a form of sudden-onset blindless in mid-life.[32] Cases of LHON typically are due to various mutations at mtDNA loci—including ND4 and ND6—that contribute to Complex I in the respiratory chain. Another such classic example is MERRF (myoclonic epilepsy and ragged red fibers), a horrible disease with shock-like muscle contractions and dementia, due to mutations in one of the mitochondrial tRNAs.[33] Table 3.4 describes these and several other human genetic disorders attributable to deleterious mutations in the mitochondrial genome itself. Mutations in mitochondrial genes have also been found in association with a wide spectrum of cancers.[34]

Many additional metabolic disorders stem from faulty nuclear-mitochondrial interactions (examples in table 3.5). These malfunctions sometimes reflect mutations in nuclear genes whose products emigrate to the mitochondrion and interact with the nucleic acid or polypeptide products of mtDNA genes; others arise when poor ATP production by mitochondria (for any reason) has negative feedback effects on metabolic systems that are nuclear encoded. Type II diabetes provides examples of both of these interrelated phenomena. For example, mutations at various nuclear genes involved in mtDNA transcription (including HNF-1α and HNF-4α, where HNF stands for hepatocype nuclear

TABLE 3.4 Examples of oft-devastating genetic disorders caused by particular mutations in mitochondrial DNA per se

---

Kearns-Sayre syndrome. A progressive disorder clinically characterized by diverse conditions including retinopathy, balance problems (ataxia), excess cerebral spinal fluid, deafness, blockage of cardiac conduction, and dementia, all appearing before the age of 20; due to spontaneous large-scale deletions in the mitochondrial genome.

Leber hereditary optic neuropathy (LHON). Characterized by atrophy of the optic nerve resulting in sudden-onset blindness in midlife; most cases result from mutations at the ND1, ND4, or ND6 genes that encode polypeptides of the NADH dehydrogenase complex; this was the first genetic disorder in humans documented to be associated with mtDNA aberrations per se.

Leigh syndrome (MILS). A devastating condition characterized by widespread degenerative complications beginning in the first year of life and often resulting in death by age three; produced by mutations in the ATPase subunit 6 gene of mtDNA (as well as by various mutations in the nuclear genome).

maternally inherited diabetes with deafness. Produced by a specific insertion mutation in a mitochondrial tRNA gene.

mitochondrial encephalo-myopathy with lactic acidosis and cerebrovascular accident episodes (MELAS). The long name of this disorder describes the medical symptoms, which include cerebral malfunctions (leading to convulsions, deafness, and dementia) beginning early in life; usually associated with point mutations in a mitochondrial tRNA gene.

myoclonic epilepsy and ragged red fibers (MERRF). A disease involving epilepsy, shock-like muscle contractions (myoclony), hearing loss, and dementia; associated with point mutations in a mitochondrial tRNA gene.

neuropathy, ataxia, and retinitis pigmentosum (NARP). Involves weakness of muscles near the trunk of the body, wobbliness, retinal disease, seizures, and delayed development; arises from mutations in the mitochondrial ATPase subunit 6 gene.

Pearson marrow pancreas syndrome (PMPS). Affects bone marrow stem cells and pancreatic function beginning in the first year of life; due to unique deletions in the mitochondrial genome.

progressive external opthalmoplegia (PEO). Symptoms include muscle weakness, exercise intolerance, droopy eyelids, and partial eye paralysis, usually appearing in adolescence or early adulthood; due to deletions of particular mtDNA sequences.

---

factor) sometimes lead to irregularities in mitochondrial gene expression that in turn precipitate Type II diabetes. And when ATP production by mitochondria is compromised for any reason (including mitochondrial mutations themselves), alterations in nuclear-encoded insulin signaling pathways often ensue, again leading to the clinical symptoms of diabetes.

Epidemiological studies of human populations in Europe and Australia suggest that the prevalence of monogenic (single-gene) mitochondrial disorders is *at least* 1 in 5,000 individuals, and could be much higher.[35] Extrapolating worldwide, this means that a minimum of 1,300,000 children and adults are seriously debilitated by genetically straightforward mutational defects in mitochondrial operations. Such tallies have surprised the medical community, which until recently viewed mitochondrial maladies as little more than rare and obscure genetic curiosities.[36]

Impressive though they are, such tallies give only the barest hint of the actual burden that mitochondrial malfunctions impose on human populations. An emerging paradigm is that many of the degenerative diseases of aging have their etiologies in mitochondria, either as deleterious mutations in the populations of mtDNA molecules themselves or as operational flaws in nuclear-mitochondrial interactions.[37]

The delayed onset and progressive course of age-related disorders certainly suggest mitochondrial involvement because the huge populations of mtDNA molecules within somatic cells gradually accumulate mutations during the lifetime of an individual and thereby precipitate energy "brownouts" in critical tissues such as the brain and heart. Growing scientific evidence indicates that mitochondrial deficiencies contribute to a wide range

TABLE 3.5 Examples of genetic disorders (and synopses of typical symptoms) *sometimes* caused by particular nuclear-gene mutations that affect mitochondrial function

---

Alpers disease (progressive degeneration of the brain; convulsions, retardation, blindness, death).

Batten disease (fatal nervous-system disorder that begins in childhood).

carnitine palmitoyltransferase I deficiency (weakness and damages from poor fat metabolism).

Friedreich ataxia (progressive damage to nervous system; gait disturbance, heart disease).

fumarase deficiency (encephalopathy due to a defect in the TCA cycle within mitochondria).

glutaric aciduria II (an illness resulting from the body's inability to metabolize fats, proteins).

hemochromatosis (iron overload leading to fatigue, abdominal pain, liver and heart disease).

hereditary spastic paraplegia (progressive stiffness of the legs, plus neurological symptoms).

Huntington disease (degeneration of brain cells, uncontrolled movements, loss of brain function).

ketone utilization disorder (difficulties from body's compromised ability to process proteins).

lethal infantile cardiomyopathy (severe muscle weakness early in life; death by 6 months of age).

Luft disease (the first mitochondrial disease reported, it involves dysfunction of muscle fibers).

malignant hyperthermia (fever, severe muscle contractions when a person receives anesthesia).

medium chain acyl CoA dehydrogenase deficiency (difficulties from altered metabolism of fats).

Menkes disease (defect in copper metabolism leading to seizures and other problems in infants).

mtDNA depletion syndrome (progressive liver failure and neurological abnormalities).

non-ketotic hyperglycinemia (seizures, mental retardation, often death due to glycine buildup).

propionic acidemia (weakness and serious problems from poor metabolism of fats and proteins).

sideroblastic anemia (iron buildup leading to anemia, possible heart disease, liver failure).
Wilson disease (copper overload leading to liver disease and neuropsychiatric problems).

---

of age-related disorders including Alzheimer disease (which currently affects about 4.5 million Americans), Parkinson's disease (affecting another one million persons in the United States), other neurodegenerative diseases such as amyotropic lateral sclerosis, heart disease, and cancers. Indeed, if aging or senescence[38] is often tied at least in part to mitochondrial dysfunctions (as now seems insuperable), then the true incidence of mitochondrial disabilities in the elderly could approach 100%. Death itself can often be interpreted as resulting from progressively severe energy brownouts that culminate in a lethal energy blackout in some critical tissue or organ.

A recent flood of technical molecular information about mitochondrial operations and dysfunctions,[39] coupled with powerful reviews about mitochondrial medicine, has catapulted mtDNA into a forefront of modern medical research. Less than a half-century has passed since Rolf Luft (a famous medical pioneer) and his colleagues diagnosed the first known patient with a mitochondrial disease,[40] and only about 20 years have passed since the first few mitochondrial diseases were detailed at the molecular level. Today, mitochondrial malfunctions are recognized as a preeminent source of human metabolic disorders.

Ludicrous Design

The serious health problems that arise from mitochondrial mutations immediately challenge any claim for omnipotent perfection in mitochondrial design. Perhaps these mutational aberrations can be viewed as unfortunate but inevitable by-products of molecular complexity. However, the intellectual challenges for Intelligent Design go much deeper than that. Considering the critical role of cellular energy production in human health and metabolic operations, why in the world would an intelligent designer have entrusted so much of the production process to a mitochondrion, given the outrageous molecular features this organelle possesses? Why would a wise designer have imbued mtDNA with some but not all of the genes necessary to carry out its metabolic role (and then put the remaining genes in the nucleus instead)? Why would a wise engineer have put *any* crucial genes in a caustic cytoplasmic environment where they are exposed routinely to high concentrations of mutagenic oxygen radicals? And why would He have dictated that the mitochondrial genetic code must differ from the nuclear genetic code, thereby precluding cross-translation between two genomes for which effective communication would seem to be highly desirable?

The puzzlement for explanations involving ID goes even further. Why would an intelligent designer have engineered mtDNA structures (such as a closed-circular genome, no introns, no junk DNA, lack of binding histones) and mtDNA operations (such as little or no genetic recombination, the production of a polygenic transcript, a limited ability to mend itself, and no self-sufficiency in transcription or translation) to differ so fundamentally from their counterpart features in the nuclear genome? In a nutshell, the underlying design of the whole mitochondrial operation seems to make no (theo)logical sense. Not only is the overall design of mtDNA suboptimal—it is downright ludicrous!

In sharp contrast, all of the mitochondrial enigmas posed above can be resolved in the light of the endosymbiotic theory of mitochondrial evolution.[41] The basic notion is that a protomitochondrion—probably a free-living purple proteobacterium—invaded (or was engulfed by) a primitive eukaryotic cell about two to three billion years ago and eventually set up a mutually beneficial symbiotic relationship with its host. The bacterium initially carried all of the genes necessary for its own survival and reproduction, but later—as the symbiotic relationship matured[42]—most of these genes either were lost or transferred to the host genome (which eventually became composed of chromosomes now housed inside the cell nucleus).

Incidentally, the transfer of copied snippets of mtDNA into the nuclear genome has continued across evolutionary time (and indeed is an ongoing process), another consequence being the widespread presence today of numts, or nuclear mitochondrial pseudogenes. Nearly 2,500 numts, ranging in size from <100 bp to 16 kb, have been identified within the human nuclear genome.[43] Many of these represent post-insertion duplications in the nucleus rather than independent insertion events,[44] and in many cases geneticists have worked out the precise evolutionary histories of the insertions and duplications through detailed DNA sequence comparisons of the extant species that share particular numts today. For example, such phylogenetic analysis indicates that the nuclear transfer of one 3,000-base-pair snippet of mtDNA occurred about 30 million years ago on the lineage leading to the Old World monkeys and hominoids (including modern humans), and that after the transfer the pseudogene became duplicated in the hominoid line.[45] Unlike functional genes in the nucleus that descended from the original bacterial endosymbiont, numts are probably worthless to cells, as gauged by their apparent lack of positional, transcriptional, or translational features that might otherwise indicate operational benefits. Like other nuclear pseudogenes (chapter 4), numts are mostly useless junk. They can also be overtly harmful, as at least one de novo numt has been documented to cause a serious human genetic developmental anomaly.[46]

A bacterial evolutionary origin for eukaryotic mitochondria effectively defuses all of the philosophical landmines described above. Extant mitochondria within our cells are simply the remnants of ancient bacterial invaders, and they still retain most of the features that continue to distinguish the genomes of most prokaryotes from those of most eukaryotes. These include tightly packed circular DNA (generally lacking introns and extensive junk sequences), rampant mutations in a genome that often proliferates clonally (without genetic recombination), and the standard production of joint transcripts from multiple genes. The mitochondrion's bacterial ancestry is supported also by molecular inspection of nucleotide sequences. The mitochondria-encoded ribosomal (r) RNA subunits of humans (and other multicellular animals) are much more similar to those in some extant bacteria than they are to the nucleotide sequences of our nuclear rRNA genes.[47]

Thus, in a very real sense, "we" (our genomes) are partly "they" (dismantled bacteria), without which we would not survive for a minute. If you do not like this reality, don't blame evolutionary biologists (who are merely the messengers of this disturbing news)—blame the non-intelligent designer himself (which from an evolutionary perspective would be selection and other natural forces). But before you fret too much about the bacterial evolutionary origins of the mitochondrial genome, save plenty of consternation for the human nuclear genome. As we will see in the next chapter, many of our nuclear DNA sequences are also microbe-like—in this case consisting of repetitive DNA sequences that look and behave much like parasitic viruses.

CHAPTER 4

Wasteful Design: Repetitive DNA Elements

Before scientists gained direct access to DNA sequences from the modern tools of molecular biology, they widely assumed that nuclear genomes were composed of sleek and efficient protein-coding genes strung together along chromosomes like tight beads on strings. But as we learned in chapter 3, structural genes themselves have complex internal structures in which the coding regions (exons) typically are like small islands in much larger hereditary rivers of noncoding introns and regulatory regions. An even bigger surprise came with the discovery that the vast majority of human DNA exists not as functional gene regions of any sort but instead consists of various classes of repetitive DNA sequences, including the decomposing corpses of deceased genes. Many repeated DNA elements inhabit the genomic outback (extensive spacer regions between genes), but others reside within functional gene regions as well. To the best of current knowledge, many if not most of these repetitive elements contribute not one iota to a person's well-being. They are well documented, however, to contribute to many health disorders.

DUPLICATE GENES

At least 4,000 protein-coding genes and other lengthy stretches of DNA (up to 200,000 base pairs in size) are present not just once but rather in small to moderate numbers of copies per genome. At least 5% of the human nuclear genome consists of such gene families in which the redundant elements (termed duplicons, or segmental duplicates) are typically more than 90% identical to one another in nucleotide sequence.[1] Members of a gene family are often arranged along one chromosome (either in tandem or separated), but sometimes they are dispersed across different chromosomes. Their close similarities of design suggest that such matched genes have arisen from one another by gene duplication processes, rather than by independent de novo craftsmanships.

The presence of duplicons often makes good functional sense, especially when large quantities of a gene product are required for cellular operations. For example, ribosomes are the active seats of protein translation (one of a cell's most fundamental tasks); accordingly, the genes that encode ribosomal RNAs are present in more than 400 copies—distributed across five pairs of human chromosomes—that collectively churn out substantial quantities of the required molecular product. Gene families also make good engineering sense when similar polypeptides are needed for related biological functions. For example, several genes in the globin family encode slightly different subunits of oxygen-carrying proteins specialized for different tissues (myoglobin in muscle, hemoglobin in blood) or different stages of life (e.g., fetal versus adult forms of hemoglobin). However, as discussed in chapter 1, in this book we are not particularly concerned with genomic features that suggest good workmanship because such features are philosophically consistent with either natural selection or intelligent causation. Our focus instead is on genomic features that defy notions of a supreme intelligence underlying biological design. Genomic flaws should in principle provide a more decisive test of whether unconscious evolutionary processes or cognitive agents have shaped our genes.

Duplicons and Their Disorders

The abundance of duplicons within a genome comes not without operational difficulties. Because duplicate genes show close sequence similarity to one another, they predispose chromosomes to pair abnormally during meiosis. This mispairing can lead to homologous recombination between various members of a gene family, generating deletions, additions, inversions, or translocations of genetic material in the resulting gametes (figure 4.1). Such genomic restructuring creates health problems in the unlucky offspring who receive such a germ-line mutation from one or both parents. Metabolic disturbances that result from duplicon-mediated genomic rearrangements typically stem from dosage imbalances (improper amounts of gene product) attributable to the presence of too many copies of a gene, too few copies,

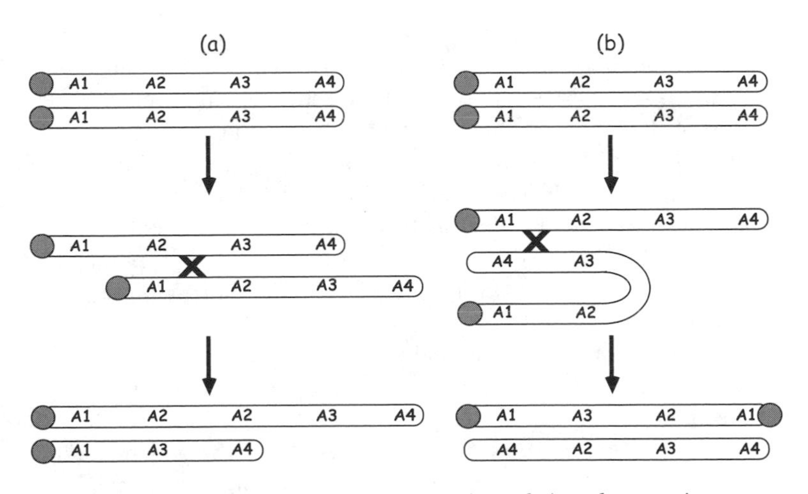

FIGURE 4.1. Duplications and deletions (panel a) and genomic rearrangements (panel b) during meiosis can emerge from physical recombination between the homologous DNA sequences of misaligned duplicons (modified from Ji, Y., E. E. Eichler, S. Schwartz, and R. D. Nicholls, 2000, Structure of chromosomal duplicons and their role in mediating human genomic disorders, *Genome Res.* **10**, 597–610). Recombination is indicated by an "X."

or to an altered orientation of particular genes relative to their regulatory regions.

Numerous metabolic disorders have been traced to duplicon-mediated recombination (examples in table 4.1). No organ system seems immune to damage; various disorders are known to affect elements of the circulatory, respiratory, hormonal, skeletal, muscular, reproductive, excretory, or nervous systems. Some genetic conditions, such as red-green color blindness, are rather benign whereas many others, such as Prader-Willi syndrome, are

TABLE 4.1 Synopses illustrating the diversity of genomic disorders for which many if not all cases are associated with duplicon-mediated rearrangements (deletions, duplications, inversions, or translocations) of particular genes

α-thalassemia. Caused by deletions in the family of four α-hemoglobin genes; symptoms range from benign (loss of one gene) to varying degrees of anemia (loss of two or three genes) to lethal before birth (loss of all four genes); highest incidences in particular populations in the Pacific and Southeast Asia.

Bartter syndrome type III. Some cases attributable to recombination between two related genes on chromosome 1 that influence chloride channels in the kidney (other cases result from point mutations at these loci); symptoms—including short stature, muscle weakness, and hypercalciuria (excessive loss of calcium into the urine)—are due to malfunctions in how renal salts are reabsorbed and homeostatic blood conditions are maintained by the kidney.

Charcot-Marie-Tooth syndrome. Caused by reciprocal rearrangements in gene sequences, on chromosome 17, probably coding for a protein involved in coating peripheral nerves with myelin; symptoms include progressive degeneration of muscles of the foot, lower leg, hand, and forearm, and loss of sensation in limbs, fingers, and toes, usually reaching full clinical expression by age 30; another genomic disorder (hereditary neuropathy with liability to pressure palsies, or HNPP) is associated with deletions in this gene region.

Congenital adrenal hyperplasia. Many cases result from recombination between a gene in the cytochrome P450 family and a related pseudogene on chromosome 6; alters how the adrenal gland produces androgen hormones; symptoms can include masculinization (development of male characteristics) of the external genitalia of females at birth.

Gaucher disease type 1. Some cases attributable to recombination between a gene for β-glucocerebrosidase and a nearby pseudogene (other cases result from point mutations within the gene itself) on chromosome 1; symptoms of this disorder of lipid storage include erosion of bone tissues and increased susceptibility to fractures.

growth hormone deficiency. Caused by deletions of a gene (*GH1*) encoding growth hormone following recombination between *GH1* and other members of the growth hormone gene family; symptoms include short stature and other difficulties associated with impaired growth and maintenance of proper amounts of body fat, muscle, and bone.

hemophilia A. Caused by inversions in a coagulation factor VIII gene following intron-mediated recombination with similar sequences on the X-chromosome; symptoms (uncontrolled bleeding) begin at infancy and range from mild to severe; affects about 1 in 5,000 males.

Hunter syndrome. Caused by an inversion of the gene encoding iduronate-2-sulfatase (*IDS*) following recombination with a nonfunctional *IDS* pseudogene on the X-chromosome; symptoms, including increased susceptibility to hernias, ear infections, runny noses, colds, coarseness of facial features, impairments of various internal organs including reduced lung capacity, and mental retardation, usually becoming increasingly noticeable after the first year of life.

neurofibromatosis type 1. Some cases are caused by recombination-mediated deletions in a duplicon that includes several genes and pseudogenes on chromosome 17; symptoms include skin disorders, nodules in the eye, bone defects, and neurological complications from tissue masses in the brain and spinal cord; the health problems stem from abnormal differentiation and migration of neural cells beginning early in embryogenesis.

Prader-Willi syndrome. Caused by deletions and rearrangements in a complex replicon on chromosome 16; affected children crave food intensely and display uncontrollable weight gain and morbid obesity often leading to lung failure, heart failure, and death.

(*continued*)

TABLE 4.1 (*continued*)

red-green color blindness. Caused by gene losses or fusions involving the red opsin locus and one or more green opsin loci located on the X-chromosome; affects approximately 8% of Caucasian males.

Smith-Magenis syndrome. A syndrome of mental retardation and congenital abnormalities due to deletions and duplications in a cluster of four genes and pseudogenes on chromosome 17; incidence about 1 in 25,000 births.

Williams syndrome. Usually caused by duplicon-mediated deletions of an elastin gene on chromosome 7; symptoms include blood vessel narrowing, farsightedness, high blood calcium levels that can cause seizures and rigid muscles; incidence approximately 1 in 20,000 births.

X-linked ichthyosis. Caused by deletions of the steroid sulfatase gene; characterized by severely dry, scaly, discolored skin, starting soon after birth; incidence about 1 in 3,000 males.

Sources: Ji, Y., E. E. Eichler, S. Schwartz, and R. D. Nicholls, 2000, Structure of chromosomal duplicons and their role in mediating human genomic disorders, *Genome Res.* 10, 597–610. See also Lupski, J. R., 1998, Genomic disorders: structural features of the genome can lead to DNA rearrangements and human disease traits, *Trends in Genet.* 14, 417–422.

severely debilitating. Many severe disorders tend not to be transmitted through families (because the afflicted seldom are able to reproduce) but instead recur in the human population from de novo mutations in paternal or maternal germ lines. Such disorders typically debilitate or kill only the individuals whose conceptions happened to involve a newly mutated sperm or egg.

Approximately 0.1% of humans who survive to birth carry a duplicon-related disability, meaning that several million people worldwide currently are afflicted by this particular subcategory of inborn metabolic errors. Many more afflicted individuals probably die in utero before their conditions are diagnosed. Clearly, humanity bears a substantial health burden from duplicon-mediated genomic malfunctions. This inescapable empirical truth is as understandable in the light of mechanistic genetic operations as it is unfathomable as the act of a loving higher intelligence.

Pseudogenes

A pseudogene is a dead gene, that is, a moribund (nonoperational) stretch of DNA that originated via duplication of a functional gene. Approximately 15,000 identifiable pseudogenes (i.e., about 0.6 pseudogenes per functional gene) reside in the human genome,[2] and this is a minimum estimate because (a) careful screening for pseudogenes across the entire genome has yet to be accomplished, and (b) pseudogenes that arose from ancient duplication events have been rendered unrecognizable by post-origin mutations. Some gene families include large numbers of pseudogenes. For the protein kinases that are important in a cell's metabolic regulation, more than 100 pseudogenes (in addition to >500 active protein kinase genes) have been identified in the human genome. Based on comparisons of nucleotide sequences of all recognizable pseudogenes and their functional gene counterparts in humans, the evolutionary half-life for pseudogene persistence is estimated to be about 40 million years.[3] After additional evolutionary time has elapsed, the decomposed remnants of former genes become no longer recognizable.

The full suite of loci for hemoglobins illustrates the evolutionary interpretation of what typically happens during the expansion of a gene family by duplicative processes. In humans, the α-hemoglobin family consists of four functional genes and three nonfunctional pseudogenes on chromosome 16, and the β-hemoglobin family consists of four genes and one pseudogene on chromosome 11 (figure 4.2). The two gene families are thought to have originated following an ancestral gene duplication event (about 450 million years ago), with subsequent duplications later generating the additional loci. Over time, members of the two families diverged such that they now encode distinct polypeptide subunits for different hemoglobin molecules with various physiological properties, each specialized for employment during a particular timeframe (embryo, fetus, or adult) in each person's development. The pseudogenes likewise arose from past gene duplications but then accumulated mutations that silenced their capacity to produce functional hemoglobins. For various hemoglobin pseudogenes, these silencing

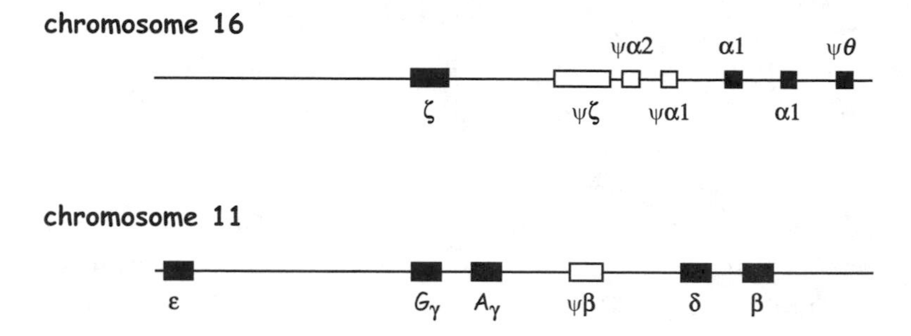

FIGURE 4.2. Genes (black boxes) and pseudogenes (open boxes) for the two gene families in the hemoglobin superfamily of loci in the human genome (modified from Li, W.-H., 1997, *Molecular Evolution*, Sinauer, Sunderland, MA; Campbell, N. A., J. B. Reece, and L. G. Mitchell, 1999, *Biology*, 5th edition, Benjamin-Cummings, Menlo Park, CA).

events are attributable to frameshift mutations, the emergence of inappropriate stop codons, and the loss of splicing sites or other regulatory elements.[4] In the globin family and elsewhere, the silencing of at least some duplicate loci makes evolutionary sense because the duplicates that remain functional continue to perform their necessary roles in the human body.

The deceased loci in the hemoglobin complex are examples of unprocessed pseudogenes, having arisen by direct DNA copying from an ancestral template gene. So too, probably, are most numts (mitochondrial pseudogenes currently housed in the nucleus; chapter 3). Processed pseudogenes or retropseudogenes, by contrast, arise when an mRNA molecule—itself the product of a gene—is reverse-transcribed into a stretch of complementary DNA (cDNA) which then reinserts into the genome. This category of pseudogenes gets its name from the fact that the mRNA molecules typically have been modified or processed before the reverse transcription event. Specifically, intron sequences usually have been spliced out and regulatory elements removed. Also, the cDNA may not cover the entire coding region of the ancestral functional gene. Thus, each resulting pseudogene is normally "dead on arrival" in the genome. Across evolutionary time, accumulated mutations serve to further decay a processed pseudogene (just as they do an unprocessed pseudogene).

Processed pseudogenes are abundant in humans and other mammals.[4,5] For example, β-actin is represented in the human genome by one functional gene and approximately 20 retropseudogenes. Comparable numbers for several other protein-coding genes and their respective deceased descendants are as follows: argininosuccinate synthetase, 1 gene versus 14 retropseudogenes; β-tubulin, 2 versus 15–20; cytochrome *c*, 2 versus 20–30; dihydrofolate reductase, 1 versus ~5; glyceraldehyde-3-phosphate dehydrogenase, 1 versus ~25; ribosomal protein L32, 1 versus ~20; and triosephosphate isomerase, 1 versus 5–6. Thus, processed pseudogenes often greatly outnumber their functional relatives. Collectively, they appear to comprise about 0.5% of the human genome. Mechanistically, the production of processed pseudogenes is closely but inadvertently connected with the activities of retrotransposable elements (see later in the chapter), which provide the enzymatic machinery necessary for generating cDNA in germ-line cells.

At face value, pseudogenes hardly seem like genomic features that would be designed by a wise engineer. Most of them lie scattered along chromosomes like useless molecular cadavers. This sentiment does not preclude the possibility that an occasional pseudogene is resuscitated such that it contributes positively to cellular operations; several instances are known or suspected in which a pseudogene formerly assumed to be genomic "junk" was later deemed to have a functional role in cells.[6] But such cases are almost certainly exceptions rather than the rule. And in any event, such examples hardly provide solid evidence for intelligent design; instead, they seem to point toward the kind of idiosyncratic genetic tinkering for which nonsentient evolutionary processes are notorious.

Pseudogenes are probably good examples of genetic elements that tend to be mildly deleterious, individually, and thus are hard to cleanse from the genome because they normally fly below the radar screen of purifying natural selection. Each pseudogene no doubt incurs a small metabolic burden on cells if for no other reason than the cellular energy and molecular materials required for its continued replication. And furthermore, even if the collective metabolic cost of legions of pseudogenes becomes substantial to cells, natural selection is not easily able to rid a population of these elements because they contribute very little to interpersonal variation in fitness. In other words, because all individuals carry huge numbers of pseudogenes, everyone is more or less equally burdened by whatever costs these elements entail, and as a result natural selection is largely impotent to remove them (unless they do more overt harm, as described below). Thus, given the relative ease of origin of pseudogenes, their persistence in genomes is understandable in the light of evolutionary reasoning.

Finally, even in the unlikely event that most pseudogenes might yet prove to have some cellular utilities that compensate for their metabolic costs of maintenance and replication within cells, the fact would remain that the homologous recombination events that pseudogenes often promote also come at considerable expense to human health. Recall, for example, that several duplicon-related genetic disorders (congenital adrenal hyperplasia, Gaucher disease, Hunter syndrome, neurofibromatosis, Smith-Magenis syndrome, and others; table 4.1) often arise from harmful recombination between a functional gene and a closely related pseudogene. Of course, such highly deleterious mutations are strongly selected against, with the inevitable costs paid in human deaths and suffering.

MICROSATELLITES AND THEIR DISORDERS

Microsatellite regions in the genome consist of modest numbers of tandem repeats of simple DNA sequences, each repeat being only two to six nucleotides long. Such tandem reiterations of short sequence are widespread in the human genome.

For example, a 6-bp sequence (TTAGGG) is found at the ends of human chromosomes in tandem arrays of as many as 5,000 to 10,000 copies. Many other microsatellite regions are composed of di-, tri-, tetra- or penta-nucleotide repeats. The dinucleotide AC, for example, is tandemly reiterated at each of more than 5,000 locations along human chromosomes.[7] Altogether, about 3% of the human genome is composed of microsatellites, which are also referred to as simple sequence repeats (SSRs).[8]

Most microsatellites have no demonstrated functional utility for a cell[9] but rather are thought to be little mishaps of DNA replication,[10] much as vocal stutters can be viewed as little mishaps in human speech. But just as stutters occasionally cause serious difficulties in verbal communication, so can microsatellites sometimes cause serious problems in human metabolism. The most famous example involves Huntington disease (HD), a fatal neurological disorder whose symptoms involve uncontrollable body movements and progressive dementia usually beginning in middle age. As detailed in chapter 2, a multidecade scientific search for the causal basis of HD eventuated in the discovery of a responsible microsatellite on the short arm of chromosome 4. Normally, 10 to 30 tandem copies of the trinucleotide CAG are found at this chromosomal position, but in HD sufferers the microsatellite region has expanded considerably (up to more than 100 copies of the CAG repeat in extreme cases). Huntington disease is a dominant disorder, meaning that even one bad copy of chromosome 4 in a heterozygous individual is sufficient to produce clinical symptoms. HD affects about one person in 10,000. Thus, in the United States alone, tens of thousands of patients and their families suffer from this terrible genetic disorder.

Huntington disease is but one of several genomic disorders recently found to be underlain by expansions of various trinucleotide repeat regions.[11] Most of these microsatellite diseases involve severe and progressive neurological impairments. Representative examples are summarized in table 4.2.

TABLE 4.2 Synopses of several genomic disorders for which many of the clinical cases have been associated with abnormal expansions of microsatellite regions in the human genome

---

dentatorubral-Pallidoluysian atrophy. A dominant disorder caused by expansion from (CAG)7–25 to (CAG)49–75 in the coding region of a gene on chromosome 12 coding for atrophin; symptoms are similar to those of Huntington disease, but may also include mental retardation in patients <20 years old.

fragile X syndrome. Caused by an expansion from (CGG)6–52 to (CGG)6–1000 in the upstream noncoding region of a gene on the X-chromosome for FMR-1 protein; this is the most frequent form of inherited mental retardation, affecting about one in every 1,500 boys and one in every 2,500 girls.

Friedreich ataxia. A recessive disorder caused by expansion from (GAA)7–34 to (GAA)>100 in an intron of a gene on chromosome 9 coding for frataxin; symptoms may include diabetes, cardiomyopathy, and degeneration of the nervous system.

Kennedy disease. A recessive disorder caused by expansion from (CAG)11–33 to (CAG)38–66 in the coding region of a gene on the X-chromosome for androgen receptor protein; usually in the third to fifth decade of life, affected individuals present with motor neuron symptoms, and muscle weakness and dysfunction.

Machado-Joseph disease. Caused by expansion from (CAG)12–37 to (CAG)61–84 in the coding region of a gene on chromosome 14 for ataxin-3; effects include progressive degeneration of parts of the brain, spinal cord, and peripheral nervous system.

myotonic dystrophy. A dominant disorder caused by expansion from (CTG)5–37 to (CTG)50–3000 in the downstream untranslated region of a gene on chromosome 19 for myotonic dystrophy protein kinase; symptoms may include progressive muscle wasting, abnormalities in heart conduction, cataracts, frontal balding, sometimes mental retardation, and testicular atrophy in males.

spinocerebellar ataxia, type 1. A dominant disorder caused by expansion from (CAG)6–39 to (CAG)41–81 in the coding region of a gene on chromosome 6 for ataxin-1; symptoms include progressive atrophy of limbs and muscles, and peripheral neuropathy.

spinocerebellar ataxia, type 10. A disease similar to the previous entry but caused in this case by expansion of a penta-nucleotide repeat from (ATTCT)ca14 to (ATTCT)>4000.

MOBILE ELEMENTS

Perhaps the most surprising genomic finding of recent decades concerns the ubiquity and abundance of mobile elements, also known as transposable elements, or, informally, "jumping genes." These stretches of DNA have—or in many cases previously had—the ability to colonize new chromosomal locations by hopping from one genomic position to another within a lineage of cells. During the process, a copy of the element is also usually retained at the original site, such that a family of mobile elements can grow inside a genome, often to astonishing numbers. Barbara McClintock discovered mobile elements in her studies of maize (corn) beginning in the 1940s, and received a Nobel Prize for her work in 1983.

Background

Almost all plants and animals harbor legions of jumping genes (or their less frisky descendants), and humans are no exception. The human genome is riddled with mobile elements, which outnumber functional protein-coding genes by approximately 100 to 1. Altogether, mobile elements constitute at least 45% of the human genome, and the true fraction is probably 75% or more if the tally were also to include (a) processed pseudogenes (see above) that originated as a byproduct of mobile element activity,[12] and (b) other intergenic DNA regions that probably originated long ago as jumping genes but are no longer identifiable as such because of post-formational mutations.

Researchers distinguish several categories of mobile elements that are well represented in the human genome (table 4.3). DNA transposons are "cut-and-paste" jumping genes that move by excision and insertion of DNA,[13] whereas retrotransposons locomote via a "copy-and-paste" mechanism of reverse transcription from an RNA intermediate.[14] Within the retrotransposon category are three principal subclasses: *LINE*s (long interspersed elements), *SINE*s (short interspersed elements), and *LTR*s (long terminal repeat elements). Each of these in turn is composed of subfami-

TABLE 4.3 Abundance data on the major classes of mobile elements in the human genome

Category	Number of copies	Fraction of genome (%)	Size of typical full element (base pairs)
transposon	294,000	2.8	2,500
retrotransposon			
LINEs	868,000	20.4	6,000
SINEs	1,558,000	13.1	300
LTRs	443,000	8.3	8,000
totals	3,163,000	44.6	

Sources: Lander, E. S. and 243 others, 2001, Initial sequencing and analysis of the human genome, *Nature* 409, 860–921; Lynch, M., 2007, *The Origins of Genome Architecture*, Sinauer, Sunderland, MA.

lies of elements. Among *LINEs*, for example, *L1* is the largest subclass. Each full-length *L1* element is approximately 6,000 base pairs long, and more than 500,000 of these elements collectively comprise about 17% of the human genome. Among the *SINEs*, *Alu* sequences constitute a major subcategory. Each *Alu* sequence is about 300 base pairs long, and approximately 1,090,000 copies of *Alu* clutter the human genome. *SVA* elements are another subfamily of *SINEs*. Each SVA element is about 1,500 base pairs in length, and about 3,000 copies reside in the human genome. Among the *LTRs*, the most abundant subclass is *HERV* (human endogenous retroviruses), which alone comprises about 8% of the human genome.

Both *LINEs* and *LTRs* (in their full-length forms) carry genes that encode proteins for reverse transcription and integration into the host genome. *SINEs* do not carry protein-coding genes but nonetheless can copy and paste by hijacking necessary enzymes (including reverse transcriptase) from other retrotransposons (especially *LINEs*). The replication of various retrotransposons is often a sloppy molecular process, so many mobile elements have lost bits and pieces that compromise their competency to code for the proteins that once enabled their own intragenomic movements. For example, reverse transcription of a *LINE* element often fails to proceed to completion, such that many of the resulting insertions are truncated and nonoperational. Fewer than 100 *LINE* elements in the human genome are thought to be active (competent to retrotranspose) today. Degenerate forms of mobile elements are called nonautonomous elements, in contradistinction to autonomous elements that are retrotranspositionally self-sufficient.

As Phylogenetic Markers

Several classes of mobile elements appear to offer superb molecular markers for assessing the phylogenetic relationships (evolutionary genealogies) of populations and species.[15] Consider SINE elements, which in suitable laboratory assays can be scored as either present or absent at any given chromosomal location in each individual or species. The number of potentially occupied chromosomal sites is huge, yet successful insertions are relatively rare in evolution, so a reasonable assumption is that each occupied site represents direct genealogical descent from a single (monophyletic, rather than polyphyletic) insertion event. Furthermore, each SINE is quite stable once inserted into a genome, so SINE absence at a given chromosomal location undoubtedly represents the original or ancestral evolutionary state. Thus, any taxa that prove upon molecular examination to share particular SINE elements almost certainly belong to an organismal clade (i.e., trace their ancestry back to a shared ancestor). Similar statements apply to various other classes of retrotransposable elements.

Phylogenetic appraisals based on mobile elements thus typically entail assaying various individuals and species for the presence versus absence of retrotransposons at each of many distinct genomic sites and compiling the resulting information into a pictorial estimate of organismal phylogeny. Results from one such molecular phylogenetic analysis, in this case involving

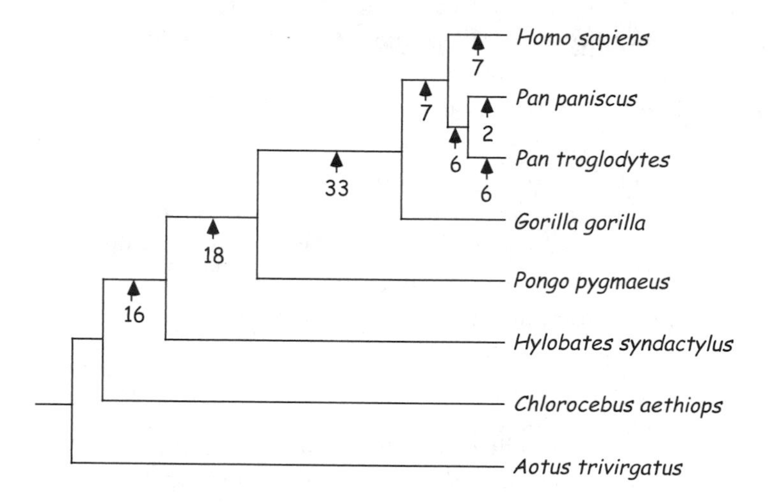

FIGURE 4.3. Phylogenetic relationships of eight primate species as deduced from the presence versus absence of numerous *Alu* elements (modified from Salem, A.-H. and nine others, 2003, Alu elements and hominid phylogenetics, *Proc. Natl. Acad. Sci. USA* 100, 12787–12791). Shown are numbers of deduced *Alu* insertion events along various branches of the evolutionary tree. The structure of the tree also agrees with a wealth of additional information from comparative morphology and genetics.

presence-absence data for approximately 100 separate *Alu* loci in eight primate species including humans, are summarized in figure 4.3. The data confirm conventional evolutionary wisdom that the two chimpanzee species (*Pan paniscus* and *Pan troglodytes*) are sister taxa related closely to *Homo sapiens*, followed in hierarchical order by *Gorilla, Pongo* (orangutan), *Hylobates* (siamang), and monkeys. Such findings about the taxonomic distributions of specific retrotransposable elements are as straightforward to interpret in the light of evolution as they are difficult to rationalize under the competing hypothesis that an intelligent designer repeatedly inserted these worthless pieces of DNA into each of multiple species not related by evolutionary descent.

What Good Are They?

Conventional scientific wisdom is that most jumping genes are intracellular genomic parasites (or, perhaps, genomic symbionts that at best are fitness-neutral to their hosts). In structure as well as operation, many mobile elements closely resemble infectious viruses (retroviruses in particular, such as the *HIV* virus responsible for AIDS), right down to the molecular details (figure 4.4). Each *LTR* mobile element is flanked in its host genome by long terminal repeat sequences, as are most retroviruses; and *LTR* mobile elements carry two of the three protein-coding genes (*gag* and *pol*) that likewise are carried by infectious retroviruses.[16] Indeed, *LTR*s are often referred to as endogenous retroviruses or retroviral-like elements, the only key distinction from infectious retroviruses being that the latter additionally carry an envelope gene whose protein product helps the virus gain cellular entry into a new host organism. Most scientists now accept

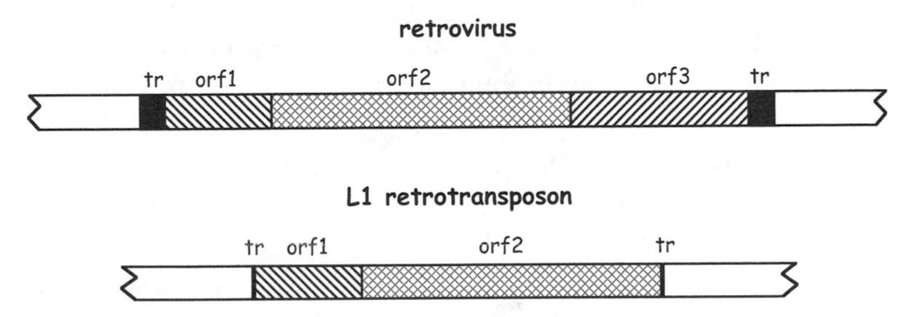

FIGURE 4.4. Simplified schematic comparison of an *L1* retrotransposable element and a typical retrovirus. Each *tr* is a terminal repeat sequence (much longer in retroviruses) that forms a boundary between the element and the host genome; *orf3* encodes an envelope protein (not possessed by most retrotransposable elements) that packages retroviruses for intercell and interhost transport; and *orf1* and *orf2* are open-reading frames containing genes encoding several proteins necessary for retrotransposition, including the keynote enzyme reverse transcriptase. The reverse transcription gene itself has seven conserved sequence blocks that are shared by all retrotransposable elements and retroviruses.

that HERVs (human endogenous retroviruses) are degenerate viruses descended from ancient infections of mammalian germ cells during evolution.

Other mobile elements, though structurally less similar to retroviruses, behave like intracellular parasites also. Some mobile elements even parasitize other genomic parasites, as for example when *SINE*s hijack the enzymatic services of active *LINE*s to transpose themselves within a host genome.

From an evolutionary perspective, the ubiquity of transposable elements relates to their proliferate nature. Some of this proliferation represents repeated historical invasions of the mammalian germ line by infectious retroviruses. Additional proliferation reflects each mobile element's activities within the genome. In a sexual species, any DNA sequence that gains a capacity in germ-line cells to generate and disperse copies of itself across chromosomal sites almost inevitably enhances its prospects for survival to succeeding generations. This would be true even if the self-proliferating DNA harms its host appreciably. Consider a hypothetical cross between an individual carrying a newly arisen mobile element and an element-free individual. From basic principles of Mendelian genetics, the mobile element would be inherited by 50–100% of the progeny depending on how efficiently it replicates to different chromosomal sites. Thus, a new jumping gene would tend to spread in the population even if it reduced the genetic fitness of each individual by as much as one-half.[17] The key to mobile element proliferation is sexual reproduction by the host organism. If the host instead were asexual, a mobile element would gain no transmission advantage by dispersing copies of itself across the clonal host genome.[18] From this perspective, mobile elements can also be regarded as sexually transmitted genomic diseases.

As with any host/parasite association, "evolutionary games" are likely to be played across time between mobile elements and their host species. In particular, host genomes are under continual selection pressure to evolve mechanisms that silence or suppress any harmful mobile element activities, and selfish mobile elements are under selection pressure to avoid such strictures. Also, some degree of self-policing by mobile elements might be expected because it is not in the element's selfish interest to harm its host. This may be especially true after a mobile element has reached high numbers in the host genome because then a selfish element's marginal fitness gains from further proliferation are diminished. The net long-term result of such co-evolutionary contests is likely to be a truce of sorts wherein mobile elements and their host genomes work out manageable ways to live together, with varying degrees of amicability.

Host organisms may even profit on occasion from their association with mobile elements, in at least two ways. First, mobile elements are powerful mutagenic agents in the genome (see below), and mutations are the ultimate source of genetic variation that is necessary for continued evolution. However, it seems doubtful that jumping genes evolved expressly for their mutagenic activities because natural selection lacks foresight (and also because many if not most random mutations are deleterious). Second, host genomes occasionally recruit formerly parasitic mobile elements into host-beneficial functions. Many mobile elements carry DNA sequences that have the capacity to regulate gene expression, and/or they may happen to insert into regulatory regions of a functional gene, so it is perhaps little wonder that host genomes sometimes have capitalized upon ("exapted") the regulatory potential latent in mobile elements for the cell's own functional purposes.[19] Such positive operational arrangements can also benefit the mobile element itself, which would then come under direct selection for evolutionary maintenance as an integral part of the host genome.

Overall, jumping genes clearly have been major drivers of human genome evolution. In the scientific literature, considerable speculation has focused on possible functional benefits that might accrue to a host organism from housing otherwise selfish and parasitic mobile elements,[20] and some of these potential benefits were discussed above. However, here we are less concerned with genomic boons than with boondoggles, and the indisputable fact remains that mobile elements cause plenty of genomic harm as well, as shown next.

Genomic Damages

Newly inserted mobile elements have the potential to cause human diseases by several mechanisms. If an element happens to land in an exon, it can disrupt the reading frame of a functional gene with disastrous consequences. If it jumps into an intron (or an intron-exon boundary), it may cause problems by altering how a gene product is spliced during RNA processing. And if a mobile element inserts into the regulatory sequence of a functional gene, it can obviously cause serious mischief. All of these possibilities fall under the category of "insertional mutagenesis," the potential for which is great. It has been estimated, for example, that an *L1* element newly inserts somewhere in the human genome in every 50–100 births, and that an *Alu* element does likewise once in approximately every 20 human births.[21] When a mobile element inserts into a host genome, it normally does so at random with respect to whether or not its impact at the landing site will harm the host. Some of the landings are soft and do little or no genomic damage. Other landings are hard and may have devastating consequences.

Another problem under the category of insertional mutagenesis is insertion-mediated deletion. When a retrotransposon lands in a functional gene, genetic instabilities are sometimes observed that result in deleted portions of the recipient locus. One example involved a new insertion of an *L1* element into a gene encoding pyruvate dehydrogenase (*PDH*). The net effect was deletion of 46,000 base pairs extending from exons 3 through 9, which in turn disrupted gene function and caused clinical symptoms (including acute neurological difficulties and severe muscle contractions) in a seven-year-old patient.[22] Other such genetic disorders likewise have been traced to genomic deletions associated with de novo insertions of *L1* elements.[23]

Apart from the difficulties associated with de novo insertions, mobile elements (or their immobile descendant sequences that previously accumulated in the human genome) can cause genomic disruptions via non-allelic homologous recombination.[24] The process is analogous to homologous recombination between members of a duplicate gene family (as discussed earlier in the section Duplicons and Their Disorders) and likewise can result in deletions or duplications of functional DNA sequences near the point of genetic recombination (see figure 4.1). Serious metabolic disorders may result. Three such genetic disorders in humans—each stemming in some cases from a recombination-mediated deletion involving *L1* elements—are glygogen storage disease, Alport Syndrome-diffuse leiomyomatosis (a syndrome of smooth-muscle tumors in the esophagus, large airways, and the female reproductive tract), and Ellis-van Creveld syndrome (involving abnormal development of the skeleton, teeth, and heart).[25] Such disorders cannot be blithely dismissed as aberrations from a genetic blueprint of optimal design. Rather, they are a direct expression of structural instabilities inherent in a genome that is riddled with troublesome repetitive elements that both invite and trigger the genetic disasters.

Most data on mobile elements (active, passive, or deceased) in humans has been gained only recently, but the list of genetic disorders associated in whole or part with these proliferate DNA sequences already is long (table 4.4). Any such list can provide only a minimum estimate of the collective toll of mobile elements on human health because most of the serious medical difficulties probably arise so early in embryonic life as to cause miscarriages that normally will remain of unknown etiology.[26] Indeed, most mobile elements are especially active in germ-line cells, so many of their deleterious effects probably register in gametic deaths.

Theological Excuses

In the context of theodicy, we must again ask why an intelligent designer with human interests at heart would have engineered a

TABLE 4.4 Human genetic disorders known to result, at least in some clinical cases, from mobile element activities. Also shown is the gene affected.

Genetic disease	Gene affected
Insertions of *LI* elements	
β-thalassaemia	*HBB*
choroideraemia	*CHM*
chronic granulomatous disease	*CYBB*
Coffin-Lowry syndrome	*RPS6KA3*
colon cancer	*APC*
fukutin	*FCMD*
haemophilia A	*F8*
haemophilia B	*F9*
pyruvate dehydrogenase deficiency	*PDHX*
X-linked cardiomyopathy	*DMD*
X-linked retinitis pigmentosa	*RP2*
Insertions of *Alu* elements	
AI porphyria	*HMBS*
adrenoleukodystrophy	*ABCD1*
Apert syndrome	*FGFR2*
aplasia anterior pituitary	*HESX1*
autoimmune lymphoproliferative syndrome	*FAS*
branchio-oto-renal syndrome	*EYA1*
breast cancer	*BRCA1, BRCA2*
cholinesterase deficiency	*BCHE*
chronic haemolytic anaemia	*P5N1*
complement deficiency	*C1NH*
Dent disease	*CLCN5*
glycerol kinase deficiency	*GK*
haemophilia A	*HEMA (VIII)*
haemophilia B	*HEMB (IX)*
hereditary desmoid disease	*APC*
hereditary nonpolyposis colorectal cancer-1	*MSH2*
hypocalciuric hypercalcaemia and hyperparathyroidism	*CASR*
Menkes' disease	*ATP7A*
Mowat-Wilson syndrome	*ZFHX1B*

neurofibromatosis	*NF1*
retinitis pigmentosa 12	*CRB1*
type 1 antithrombin deficiency	*SERPINC1*
Walker Warburg syndrome	*POMT1*
X-linked agammaglobulinaemia	*BTK*
X-linked hyper-IgM	*CD40LG*
X-linked severe combined immunodeficiency	*IL2RG*
Insertions of *SVA* elements	
autosomal recessive hypercholesterolemia	*ARH*
Fukuyama congenital muscular dystrophy	*FCMD*
hereditary elliptocytosis and hereditary pyropoikilocytosis	*SPTA1*
X-linked agammaglobulinemia	*BTK*
Recombinations among *Alu* elements	
acute myeloid leukaemia	*MLL*
α-thalassaemia	*α-globin*
angioneurotic oedema	*G1-inhibitor*
B-cell lymphoma	*RBL1*
chronic granulamatous disease	*NCF1*
complement component 3 deficiency	*C3*
Duchenne muscular dystrophy	*DMD*
Ehlers-Danlos syndrome type VI	*PLOD*
Ewing sarcoma	*EWSR1*
Fabry disease	*GLA*
familial breast cancer	*BRCA1*
Fanconi anaemia	*FAA*
Glanzmann thrombasthenia	*ITGAZB*
gliomas	*RB1*
glycogen storage disease VIII	*PHKA2*
haem oxygenase-1 deficiency	*HMOX-1*
hereditary nonpolyposis colorectal cancer	*MLH1*
hypercholesterolaemia	*LDLR*
hypo-betalipoproteinaemia	*APOB*
insulin-independent diabetes mellitus	*INSR*
Lesch-Nyhan syndrome	*HPRT*
mucopolysaccharidosis IVA	*GALNS*

(*continued*)

TABLE 4.4 (*continued*)

Sandhoff disease	*HEXB*
severe combined immunodeficiency	*ADA*
Tay-Sachs disease	*HEXA*
thrombophilia	*SERPINC1*

Sources: Hedges, D. J. and P. L. Deininger, 2007, Inviting instability: transposable elements, double-strand breaks, and the maintenance of genomic integrity, *Mutation Res.* 616, 46–59; Prak, E. T. L. and H. H. Kazazian, 2000, Mobile elements and the human genome, *Nature Rev. Genet.* 1:134–144.

genome grotesquely infested with parasitic elements. There is no scientific debate that active or deceased mobile elements are pervasive in the human genome (indeed, the human genome is composed mostly of "their" DNA rather than "ours," such that in overall genetic makeup "we" are primarily "them"). There is also no debate that the actions of mobile elements can and often do disrupt genomic operations in ways that cause deaths and human suffering. From a design perspective, the compensatory feature (if any) would seem to be that DNA sequences from formerly mobile elements occasionally are co-opted into host-beneficial services.[27] Perhaps an intelligent designer had such second-order benefits for us in mind when she created mobile elements to begin with. But then why would she have left the happenstantial "exaptation" processes[28] to the vagaries of evolution, as opposed to ensconcing the human genome with consistently beneficial structures right from the get-go?

Once again, transposable elements hardly reflect the kind of biological engineering that most proponents of Creation Science and Intelligent Design have in mind when they speak of the complexity and beauty of God's works. At best, a genomic jungle infested with parasitic DNA sequences would seem to be a dangerous, ill-conceived, and highly circuitous route by which to engineer occasional genomic advantages. However, from a more relaxed theological viewpoint (far less extreme than that characteristic of Intelligent Design), perhaps jumping genes could be considered just part of the grand but otherwise natural evolutionary process though which God creates.

Evolutionary Excuses

From an evolutionary perspective, no insuperable dilemma exists for rationalizing the presence of parasitic jumping genes: natural selection is just as devoted to the parasite as it is to the host. In any sexually reproducing species such as *Homo sapiens*, recombinational processes (meiosis and syngamy) in each generation ensure that different pieces of unlinked DNA (i.e., different loci) have noncoincident transmission histories and, thus, quasi-independent evolutionary fates. This means, in effect, that evolutionary forces including natural selection can scrutinize individual loci as partially separable units and thereby influence the dynamics of DNA sequences in some ways that would otherwise be impossible. For example, once a segment of DNA has gained residence in a sexual genome, it can compete as well as collaborate with other loci (functional genes included) for successful transmission to succeeding generations. This opens a window of opportunity for the evolution of selfish (in addition to cooperative) behaviors among different DNA segments within a sexual genome. More than 30 years ago, Richard Dawkins was among the first to popularize the notion of the selfish gene,[29] and the concept is now part of mainstream wisdom in evolutionary genetics.

The evolutionary consequences of such intragenomic interactions are profound, as documented in a recent treatise by evolutionary biologists Austin Burt and Robert Trivers.[30] These authors canvass the many ways that loci in sexual species have discovered to spread and persist in populations by selfish means, that is, without contributing positively to organismal fitness. In theory, any piece of DNA could increase the proportion of gametes (and thereby zygotes of the next generation) in which it is represented by employing any of at least three genuinely selfish tactics: *interference*, in which the selfish allele disrupts or sabotages the transmission of alternate alleles at the same locus; *gonotaxis*, by which selfish DNA moves preferentially toward the germ line; and *overreplication*, in which a segment of selfish DNA biases its intergenerational transmission by getting itself replicated more often than other loci in the same host. Each such selfish tactic imposes (by definition) a negative fitness cost on the host organism, but a selfish element nonetheless can spread in a host population if on balance it proliferates or "drives" more than it decreases the fitness of its host.

Loci tightly linked to the selfish driver may also benefit, by hitchhiking and perhaps also by evolving mechanisms that enable it to share in the driving effort. By contrast, the genetic fitness of loci unlinked to the driver DNA is often harmed by the actions of selfish genes, such that much of the host genome comes naturally under selective pressure to police the selfish drivers and get them off the hereditary road.[31] This might transpire via the evolution of genomic mechanisms that ameliorate or suppress the selfish gene's bad-driving habits, or by the evolutionary recruitment of a selfish gene into host-beneficial services (as described above). All of these co-evolutionary phenomena are empirically well documented, in diverse guises, in various sexually reproducing species.

Mobile elements generally fall into the overreplication category of selfish behavior. In other words, they have come to be present in such great abundance because they have found mechanistic ways to proliferate within host genomes. Another contributing factor probably relates to the long-term effective size of the host population,[32] which is (or has been) typically rather small for most sexual species including humans. In theory, the simple requirement for expansion of a family of mobile elements is that the rate of element proliferation exceeds the rate of element removal by natural selection. The latter is influenced not only by the magnitude of the element's negative impact on host fitness but also by the effective size of the host population (because the smaller the host population, the more oblivious natural selection becomes to a given magnitude of genomic damage, and the less effective it becomes in removing or silencing the offensive agents). Even moderately harmful mobile elements thus stand a chance of proliferating in a host species with small effective population size, until perhaps the cumulative harm of these elements generates a negative fitness signal that becomes audible to natural selection above the background noise.

From an evolutionary vantage, essentially it's as simple as that. There is no need to invoke non-natural causation for mobile elements any more than there is a need to invoke non-natural processes to explain the spread of infectious viruses in susceptible host populations. Indeed, who would wish to invoke intelligent design to explain the activities of either mobile elements or disease-causing viruses? Or, almost equivalently, who would therefore wish to invoke intelligent design to explain the genesis of at least this 50% or more of the total human genome?

Toward the end of his illustrious career in biology, J. B. S. Haldane was asked by a news reporter what he had learned from his years of evolutionary study. With tongue only partially in cheek, Haldane responded, "The Creator must have had an undue fondness for beetles" (because He made so many species of them). If Haldane were alive and asked that same question today, in the genomics era, he might well respond, "The Creator must have had an undue fondness for mobile elements."

CHAPTER 5

Intelligent or Non-Intelligent Design?

In *Darwin's Black Box: The Biochemical Challenge to Evolution*, biochemist Michael Behe wrote (p. x): "Modern science has learned that, ultimately, life is a molecular phenomenon: All organisms are made of molecules that act as the nuts and bolts, gears and pulleys of biological systems"; and (p. 4) "Life is lived in the details, and it is molecules that handle life's details." Behe then raised the question (p. 4), "Can it [Darwin's idea] explain life's foundation?" Behe's answer was a firm "no": "Although Darwin's mechanism—natural selection working on variation—might explain many things, I do not believe it explains molecular life" (p. 5).[1] By default, Behe thereby invoked intelligent design to account for life's biochemical complexity. Behe's prose is engaging and disarming, and his proclaimed respect for empirical evidence and logical argument in addressing causation in biology are to be applauded. I am sympathetic to many subsidiary points in Behe's book, but I think the author is flat wrong in his overarching thesis that the molecular details of biological complexity register intelligent design.

In the preceding chapters we have delved into many structural and operational intricacies of the human genome, the ultimate biochemical foundation of each person's physical existence. Many of these molecular arcana were illuminated only recently (some of them only after the publication of Behe's book in 1996), and many were unanticipated—even shocking—to the scientific community. In this concluding chapter, we will reflect on these remarkable genomic discoveries and consider their philosophical implications with regard to intelligent versus non-intelligent design. I will use Behe's 1996 treatise as a touchstone for discussion because it remains the preeminent book-length endorsement of ID from a professional molecular biologist.

RAMPANT MOLECULAR COMPLEXITY

Behe states (p. 173), "Darwin never imagined the exquisitely profound complexity that exists even at the most basic levels of life." Biotic complexity is indeed exquisitely profound at the molecular level, if "exquisite" is interpreted as "compelling high admiration" (one dictionary definition) rather than implying absolute perfection according to a hypothetical ideal. Today, in the era of high-throughput DNA sequencing and genomic-level analysis, there is no dispute that molecular structures and operations in the human genome are fantastically complex, far beyond what Darwin might have imagined from the phenotypic information available in his time.

Each human genome consists of more than three billion nucleotide pairs, only a tiny fraction (<1%) of which codes for amino acids that comprise the tens of thousands of functional proteins supporting the scores of complicated biochemical pathways that underlie metabolism in each of a person's trillions of somatic cells. Protein-specifying loci themselves are subdivided into polypeptide-coding exons interspersed with 30-fold larger (on average) introns, with the latter being scrupulously removed by the cell from each transcript before various exon sequences are spliced together in what are usually proper combinations. Many protein-coding loci have become duplicated repeatedly to yield families of related genes, some of whose members later became functionally deceased but whose cadavers still litter the human genome. Many other such pseudogenes have decomposed so thoroughly as to now be mere hummus in the genomic substrate.

Flanking each protein-coding region are batteries of regulatory DNA sequences that interact with complex arrays of activator and repressor proteins to help govern transcription, which is merely one of about a dozen key stages of protein production and utilization at which cells exert regulatory controls. Other transcription factors, modulators, and modifiers of gene expression are encoded by trans-operating loci that are unlinked to the genes whose protein products and biochemical pathways they regulate. These include vast arrays of loci encoding protein kinases, microRNA molecules, and other regulatory modulators.

Perhaps most remarkably, protein-coding loci have proved to be mere tiny islands of DNA in the vastly larger river of intergenic spacer sequence. Much of the latter is composed of mobile elements that closely resemble viruses, both in their physical structures and in terms of their cell-parasitic evolutionary behaviors. Indeed, at least 45% (and probably much more) of the human genome derives from mobile elements that have proliferated selfishly or parasitically across the chromosomes. Some of these mobile elements remain active today, but many more have become debilitated and now lie dormant. Perhaps not surprisingly, some of these impaired sequences have been recruited secondarily into host-beneficial services, such as helping the cell to regulate gene expression.

Adding further to human genomic complexity is the fact that in each generation of sexual reproduction, our genes are routinely mutated, segregated, and reassorted into novel multilocus combinations, each of which is unique in human history and therefore has never before been field-tested for proper performance. Sexual reproduction has profound implications because genes that routinely are shuffled in a hereditary lineage ineluctably assume partially independent evolutionary fates.[2] In effect, natural selection and genetic drift then can scrutinize genes as quasi-separate or individualized units across the generations and thereby influence the dynamics of genomes in ways that otherwise would be impossible.

Most notably, the ever-shifting genomic alliances fated by genetic recombination in sexual taxa can yield conflicts of interest among otherwise collaborative genes and can open windows of opportunity for the evolution of selfish or parasitic (in addition to cooperative) behaviors for unlinked loci. Selfishness by an endogenous piece of DNA means persisting and perhaps proliferating copies of itself in a host population without enhancing the immediate well-being of the host organism.[3] (Selfishness here does not imply any awareness by the element of the consequences of its actions; a parasite need not be conscious of the harm it does its host.) Mobile elements provide quintessential examples of selfish DNA sequences in sexual species.

The behavior of mobile elements contrasts with evolutionary expectations for genes in strictly clonal lineages (if such exist). All genes in a clonal genome are co-transmitted, so in theory the fates of all loci would be coupled and over evolutionary time they should tend to evolve "all-for-one-and-one-for-all" proclivities. In other words, any gene in a strictly clonal reproducer could improve its prospects for representation in future generations only by helping (or at least not strongly impairing) its host's survival and reproduction. In any genuinely asexual lineage, therefore, opportunities simply would not arise for the evolution of selfish renegade genes. This consideration may help to explain the relative genomic simplicity and the relative paucity of transposable elements in unicellular prokaryotes (many of which show tendencies for clonal reproduction, at least occasionally) when contrasted with the high genomic complexity and the profusion of transposable elements that characterize most multicellular eukaryotes (most of which are obligately sexual). However, several other biological considerations undoubtedly contribute to the relative simplicity of prokaryotic genomes because prokaryotes differ from eukaryotes in many features apart from clonal propensity.[4] An absence of effective genetic recombination may also help to explain the genomic streamlining and the absence of mobile elements in animal mitochondrial DNA, which is descended from bacterial ancestors in the distant past and remains clonally transmitted in most animal species today.

For the nuclear genomes of sexual creatures such as humans, the weight of empirical evidence for molecular complexity has compelled scientists to abandon simplistic genomic images (such as the former beads-on-a-string model for genes along chromosomes) and to contemplate more sophisticated metaphors for genomic operations.[5] One such metaphor likens each sexual genome to a social collective in which different DNA sequences (i.e., different genes or sets of unlinked loci) display fine divisions of labor and functional collaborations yet maintain partial autonomies of fate that can result in occasional conflicts of evolutionary interest. Under this metaphor, many categories of gene interactions across evolutionary time roughly mirror those of humans bound in social units such as communes. These behaviors include collaborative efforts among the genes, but also cheating; aggregate actions, but also personal opportunism by specific loci; group alliances, but also intragenomic conflicts; and parliamentary needs opposed by egoistic tendencies of particular pieces of DNA.

Within each such genomic community of nucleic acid sequences, protein-coding genes in their nonmutated states tend to be good-citizen loci, each contributing in some small but perhaps crucial way to an organism's survival and reproduction. The same is true for many other RNA-yielding loci, such as those that encode ribosomal RNA subunits, tRNAs, and miRNAs. Actively proliferating transposable elements, by contrast, are prime examples of outlaws within the genomic community, each proliferating across the genome and thereby enhancing its own prospects for transmission to future generations almost regardless of any negative consequences for the host organism. Of course, such evolutionary selfishness does not preclude the possibility that some mobile elements are evolutionarily recruited, secondarily, into host-beneficial services (examples are described in chapter 4).

In this emerging metaphor of the genomic lineage as a miniature social collective of interacting loci, pseudogenes are the cell's quintessential indigents. These deadbeats mostly just lie about the genome, absorbing societal resources without generating meaningful returns to the cell. Occasionally, a pseudogene becomes rehabilitated and resumes a useful life in genomic society (chapter 4). A pseudogene that has returned to a cell's good graces may play a role resembling its former function, but sometimes its new profession may be quite different from the original (especially after additional rounds of evolutionary tinkering by mutation and natural selection). Nevertheless, most pseudogenes are probably useless residua that take up space and waste cellular resources without offering immediate useful payback to the remainder of genomic society.

Another metaphor might envision each genome as a miniature ecosystem within an extended (multigeneration) cellular lineage. Each operational gene occupies a particular functional niche, and different DNA sequences will have evolved elaborate interactions—including mutualistic and parasitic-like behaviors—normally associated with various suites of biological species in natural macro-ecosystems. For example, standard good-citizen genes (e.g., most protein-coding loci) are mutualistically interdependent as they collaborate to produce a viable organism, whereas transposable elements are like mini-parasites within a genomic micro-ecosystem. This metaphor falls short, however, in the sense that different loci within a cellular micro-ecosystem are often knit together more tightly (both structurally and functionally) than are many species in a macro-ecosystem.

Regardless of the adequacy of these or other such genomic metaphors, the broader point is that recent molecular analyses have unveiled complexities in genomic structure and function that far surpass what most biologists (and theologians) in times past might ever have envisioned.

REDUCIBLE COMPLEXITY

To Behe, a biochemical or other organismal trait that most clearly evidences intelligent design is irreducibly complex, meaning that it experiences a complete loss of function upon the removal of any of its components. In Behe's words (p. 39),

> By irreducibly complex I mean a single system composed of several well-matched, interacting parts that contribute to the basic function, wherein the removal of any one of the parts causes the system to effectively cease functioning. Furthermore (p. 39), an irreducibly complex system cannot be produced directly (that is, by continuously improving the initial function, which continues to work by the same mechanism) by slight, successive modifications of a precursor system, because any precursor to an irreducibly complex system that is missing a part is by definition nonfunctional.

Behe's contentions notwithstanding, I fail to see how the dependency of a functionally complex trait upon each of its constituent parts necessarily eliminates historical etiologies for such biological traits. Consider figure 5.1, which shows a hypothetical sequence of historical events culminating in the evolutionary emergence of refined vision. During each of the five steps in this historical progression, a new "gene" (i.e., a novel genetic capability) was added to the genome, resulting in a vision improvement. At the right side of this diagram, the eye of any extant species (such as *Homo sapiens*) would have the capacity to see colors as well as shades of gray, and that eye would also offer sharp focus for refined perceptions of shape and movement. Suppose now that a person inherits a deleterious mutation in gene 1 that completely blocks her capacity to detect light. Sadly, the other four vision genes would be of no avail in rescuing this woman's sight because the blockage of light would fully preclude her capacity to detect (much less focus on) shades of gray, colors, shapes, or motions. Furthermore, depending in part on how the remaining genes are metabolically linked and regulated, deleterious mutations at any of the other genes could in principle likewise result in the loss or severe impairment of vision. This is a straightforward example of how a biological trait that is irreducibly complex (under Behe's definition) nevertheless could have arisen gradually and cumulatively via standard evolutionary processes.

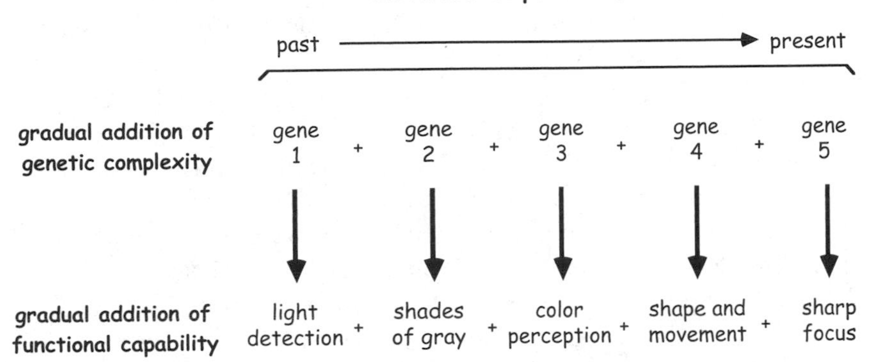

FIGURE 5.1. Diagrammatic representation of how an "irreducibly complex" trait (such as the vertebrate eye) can in principle evolve gradually, without the invocation of intelligent design (see text for details).

The actual evolution of the vertebrate eye (a quintessentially complex trait) proceeded stepwise, probably in a fashion at least qualitatively similar to that depicted in figure 5.1. A similar gradual transformation of vision is also implicated for mollusks (invertebrate animals that include snails and their allies), which collectively show a succession of photoreceptor organs ranging from the simple light-gathering pigment spots of the limpet to the invaginated optic cups of the nautilus, and to the astonishingly sophisticated eyes (complete with iris, refractive lens, cornea, retina, and optic nerve) of the octopus. From similar kinds of comparative analyses in the context of phylogeny, evolutionary biologists likewise have been able to reconstruct the probable succession of events underlying many other sophisticated biological systems—such as the biochemistry of blood clotting[6]—that proponents of ID often have claimed to be irreducibly complex. Although many such traits indeed have proved to be astonishingly complex, such complexity typically appears to be scientifically reducible (with sufficient evolutionary detective work).

Another of Behe's prime examples of irreducible complexity is the bacterium's tail (flagellum), which consists of many working parts that Behe claims could not have arisen stepwise, by evolution. Due no doubt to the emphasis that Behe placed on the flagellum, biologists quickly set to work to examine this structure more closely. For example, Renyi Liu and Howard Ochman reported the identification of more than 50 genes that encode structural components of the flagellum.[7] They then investigated all flagellar proteins encoded in the full genomic sequences of 41 bacterial species representing 11 phyla, and demonstrated phylogenetic relationships suggestive of the evolutionary order in which these structural components of the flagellum arose. The authors concluded, "core components of the bacterial flagellum originated through the successive duplication and modification of a few, or perhaps even a single, precursor gene." Furthermore, some of the components of the flagellum appear to reflect exaptations that evolved originally to perform other roles (related to bacterial secretion rather than mobility). Other scientific considerations of flagellae similarly have reached the conclusion that these elaborate bacterial structures are entirely consistent with step-by-step evolutionary scenarios.[8]

In chapter 7 of *Darwin's Black Box,* Behe develops an extended metaphor in which the probability of gradualistic step-by-step evolution of a complex trait is likened to the probability of a blind groundhog successfully crossing a multilane expressway of speeding cars and 18-wheelers. The poor little animal would have no chance and would end up as Road Kill (the title of Behe's chapter). The metaphor seems compelling at face value, but it could be modified in realistic ways that make it more relevant to the biological problem at hand. First, suppose that many groundhogs (rather than just one) initiate the journey, and suppose further that the highway has median strips or islands between lanes that provide safe havens for the traveling rodents.[9] Relatively few animals might make it to the first sanctuary, but the fortunate survivors could then reproduce sufficiently to reconstitute the population before the next generation of groundhogs proceeded across lane 2. And so the process would continue, generation after generation across successive lanes and dividers, until some distant groundhog descendants eventually might reach the far side. The stepwise journey—which is far from implausible—would be quite analogous to the stepwise evolution, as described in chapter 1, of bacterial resistance to multiple antibiotics.

At least three additional factors could further facilitate the groundhogs' collective trip. First, the speed and volume of motorized traffic probably varies throughout the day as well as seasonally, and periods with light traffic could be deemed analogous to times of relaxed selection on groundhogs (or perhaps to situations when relatively simple genetic changes underlie the relevant evolutionary transitions). Second, the selection pressures imposed by cars and trucks would ensure that the groundhogs who survive each round carry nonrandom sets of genes with respect to features such as speed, agility, or eyesight, so adaptive traits of these sorts would tend to increase in frequency across the generations. Third, the groundhogs might possess various exaptations (chapter 4) or preadaptations for traffic. Some of these might be so obvious as to be overlooked. For example, by having evolved natural instincts to fear and evade large mobile predators such as mountain lions and coyotes, groundhogs are evolutionarily preadapted to some extent for the behavioral avoidance of other large moving objects (including cars and trucks).

The concept of exaptation—which IDers routinely neglect when discussing irreducibly complex traits—warrants elaboration. The concept of a Darwinian exaptation comes from the realization that a trait originally engaged in one biological operation or suite of functions sometimes assumes a new role during the evolutionary process. A classic example—mentioned in chapter 1—involves the three bones (mallus, incus, and stapes) of the mammalian inner ear, which now play a key role in human hearing but that much earlier in vertebrate evolution were components of the jaw apparatus in our distant reptilian ancestors. Another familiar example is our opposable thumb, which served our more recent ancestors well in climbing trees but now finds utility in helping us to manipulate screwdrivers and iPods.

The phenomenon of exaptation is also common—indeed rampant—at the molecular level. It is evidenced, for example, whenever a mobile element in effect is lassoed by the host genome and domesticated into host-beneficial services, such as providing a new mechanism for regulating gene transcription. It is registered by the presence of introns that now play key roles in alternative splicing and polypeptide assembly, but which in their former genomic lives may have been constituents of mobile elements or perhaps pseudogenes. The exaptation phenomenon is also evidenced by any gene duplication event that has led to descendant loci with novel cellular functions. Empirical cases are legion. Two such examples were described in detail in earlier chapters: globin genes, which arose by a series of gene duplication events in vertebrate evolution and subsequently became specialized for delivering oxygen to different tissues or at different stages of life; and the huge protein kinase family of genes, which likewise diversified following successive duplication events into a plethora of specialized regulatory roles in various cellular metabolic processes.

With respect to their magnitude of functional impact, biological alterations that originated as exaptations can range from the minute to the profound. In humans, the fetal versus adult forms of β-hemoglobin, which trace their ancestry to a gene duplication event, exemplify a relatively small (but still important) adaptive refinement. An example near the other end of the magnitude scale involves a molecular apparatus—the mitochondrion—by which eukaryotic cells generate most of their biochemical energy. This cytoplasmic organelle is now fully integrated into the molecular workings of our cells, but the evolutionary precursor of its genome originated early in the history of life during an endosymbiotic merger of unrelated microbes. Thus, this maternally inherited genome can be interpreted—almost literally—as one of the grandmothers of all exaptations.

Many additional examples of molecular exaptation have been unearthed by careful evolutionary-genetic dissections of complex adaptations. The results of such scientific investigations notwithstanding, advocates for ID will always be able to nominate some other complex biological trait, not yet well investigated from a molecular or evolutionary-genetic perspective, as a possible next candidate for intelligent design. Like repeated false cries of "wolf" or "fire," such nominations at some point will lose most of their force to an objective listener. They will probably continue to be important in creationist manifesto, however. As noted more than 40 years ago by the eminent evolutionary biologist Theodosius Dobzhansky,[10]

> There are people...to whom the gaps in our understanding of nature are pleasing for a different reason. These people hope that the gaps will be permanent, and that what is unexplained will also remain inexplicable. By a curious twist of reasoning, what is unexplained is then assumed to be the realm of divine activity. The historical odds are all against the God-of-the-gaps being able to retain these shelters in perpetuity. There is nothing, however, that can satisfy the type of mind which refuses to accept this testimony of historical experience.

In other words, no matter how compelling the empirical evidence for natural selection, some people will remain, for ideological or other reasons, persuaded that intelligent design provides a better explanation than evolution for complex adaptations; ergo our preoccupation in this book with genomic flaws and biochemical systems that often fail abjectly, and for which ID cannot be invoked so easily. In general, proponents of ID tend to place the onus on scientists to explain every fine detail in the seeming perfection of complex biological features. By focusing here instead on molecular imperfections, I hope to shift at least some of the burden of etiological evidence and logic onto creationists' shoulders.

Additionally, in the current context of exploring the origins of complex biological traits, a focus on molecular flaws is further justified by the fact that both scientists and theologians have tended to underemphasize molecular blemishes compared to molecular elegances. Biologists sometimes have neglected biochemical imperfections in their zeal to document the creative power of natural selection, whereas natural theologians routinely overlook molecular flaws in their quest to prove intelligent design.

INEVITABLE IMPERFECTION

Behe promulgated the notion of irreducible complexity at the biochemical level in his objection to natural evolutionary causation, stating that (p. 22) "biochemistry offers a Lilliputian challenge to Darwin." However, in the recent empirical light of rampant genomic defects and biochemical breakdowns, it might be said with even greater force that the findings of molecular biology offer a Gargantuan challenge to intelligent design. From scientific evidence gathered during the last century, and especially within recent decades, we now understand that the human genome and the metabolic processes it underwrites are riddled with structural and operational deficiencies.[11] These defects register not only as deleterious mutational departures from some hypothetical genomic ideal, but also as universal architectural flaws in the standard genomes themselves. In other words, "inevitable imperfection" at the molecular level now seems to be far better documented, scientifically, than does irreducible complexity.[12] These findings mean that the age-old theodicy challenge, traditionally motivated by obvious imperfections at the gross levels of human morphology and behavior, now can be seen to extend into the innermost molecular sanctum of our physical being.

Mutational Glitches

This much is indisputable: gene-based metabolic disorders are widespread in human populations. Ever since the pioneering work of Sir Archibald Garrod in the early 1900s, these have been termed inborn errors of metabolism, and rightly so. They are biochemical blunders—cellular malfunctions directly attributable to endogenous frailties of the human genome. These genetically based errors cause tremendous human pain and suffering. They afflict saints and sinners alike, the young (including embryos and fetuses) and the old, the wise and the ignorant, the rich and the poor. Indeed, metabolic malfunctions are ubiquitous and deadly.

If some other source of mortality (such as a bullet or a car accident) does not intervene first, a metabolic disorder sooner or later will kill each of us. We may not like this biological state of affairs, but it is the indisputable reality.

Thanks to the laborious efforts of geneticists and biochemists, we have gained detailed molecular knowledge about the proximate etiologies of hundreds of human metabolic diseases. Each disorder typically results from a deleterious mutation in one or another of the thousands of structural and regulatory genes that populate our genomes. In many cases, the mutant locus (or loci) for each genetic disorder now can be specified precisely, as can its laboratory diagnosis and its operational consequences for a patient's tissues and organs. Healthcare professionals and their clientele currently have access to many thousands-fold more information on human inborn errors of metabolism than Sir Archibald Garrod managed to accumulate across his lifetime. As phrased by Behe (p. 232), "Over the past four decades modern biochemistry has uncovered the secrets of the cell." Although many of those secrets are beautiful and awesome, others can only be described as ugly and contemptible.

Inherent Design Flaws

Apart from its proclivity for being debilitated by de novo mutations, the human genome in its "normal" (nonmutated) states exhibits profound design flaws. Such imperfections, detailed at length in the preceding chapters, are not unexpected if insentient evolutionary forces engineered the genome. Evolution is a sloppy tinkering process, constrained by happenstance and historical precedence and guided with respect to adaptations by a mindless directive agent (natural selection). Evolution's biological products, including the human genome, abundantly reflect this complicated, history-laden reality. Nevertheless, we can speculate here on what the genome might be like had it been designed and built by an intelligent engineer.

Many architectural routes to perfection or near-perfection in the human genome might be envisioned, but for the sake of argument let us assume that a Creator God had valid reasons for choosing nucleic acids and polypeptides (rather than some other biochemical substances) as the molecular foundations for life. Notwithstanding that restriction, any omnipotent designer with human interests at heart surely could have engineered a genome that is far better suited to most people's health and well-being.

The following are some of the molecular features that a wise and beneficent engineer probably should and almost certainly could have incorporated into the human genome: (a) more efficient repair mechanisms for DNA, RNA, and proteins, cleverly targeted on damaged molecules that actually harm human health; (b) a freedom from mobile elements and other stretches of selfish DNA that proliferate at host expense (or at best at host indifference); (c) a freedom from pseudogenes, microsatellites, degenerate mobile elements, and other operationally useless pieces of DNA that additionally squander cellular resources; (d) more streamlined and efficient protein-coding genes, not split into segments by introns,[13] and regulated in a more straightforward and less error-prone fashion; (e) a relief from the intergenic strife precipitated by genetic imprinting and by many other factors related ultimately to sexual reproduction; and (f) a mitochondrial genome—if the cell needs a separate one at all, not integrated into the nucleus[14]—that encodes a full suite (rather than only a tiny subset) of the polypeptides and RNA molecules necessary for cellular energy production. This list of genomic improvements could easily go on.

One might expect an intelligent engineer also to have added some beautiful finishing touches to her product, if for no other reason than pride of workmanship. For example, the designer could have placed functionally related genes (such as those belonging to each specified biochemical pathway) adjacent to one another in the human genome, and vastly simplified and streamlined their regulatory control, for example, by arranging that they be processed jointly or coordinately. Even more fundamentally, she could have engineered the genome to be mutation free, thereby eliminating many overt genetic disorders and precluding requirements for the metabolically expensive and complicated repair apparatuses that our cells currently deploy.

An ardent proponent of intelligent design might suggest that such genomic refinements are for some reason physically or biochemically impossible for life as we know it. But such a stance would be demonstrably incorrect. Many of the features listed above (a notable exception being a freedom from deleterious mutations) are in fact characteristic of prokaryotic genomes, and several of them also characterize human mitochondrial DNA.[15] If an intelligent designer has shown such care in constructing the genomes of lowly bacteria, why would she not have crafted the human nuclear genome with at least a comparable level of loving attentiveness?

An ID proponent might rejoin that humans are so complex (compared to bacteria, for example) that ramping up the bacterial genomic designs to the scale required for proper human development was technically impossible for the Creator God. Constructing a human probably requires many thousands more proteins than does constructing a bacterium. Thus, some theologians might argue, the intelligent designer acted wisely when she imbued the human genome with introns (necessary for alternative splicing), complex regulatory controls (for somatic differentiation in development), mutations (that despite being incessant are also constrained by cellular repair mechanisms), and a multigenerational adaptive flexibility that mobile elements, pseudogenes, and other noncoding pieces of DNA can further help to provide. However, this kind of argument would be closer to an evolutionary than to a strict Intelligent Designer's explanation for life. It would imply not only that the intelligent agent was less than omnipotent but also that her primary role was to set into operation natural evolutionary forces—including mutation, selection, genetic drift, and sexually mediated genetic recombination during reproduction—that ever since have been responsible for life's diversification (including organic flaws as well as useful adaptations). This sounds far more like the workings of Mother Nature than it does the workings of an Abrahamic Creator God who can invoke miracles at will. Such evolutionary-like explanations are probably compatible with many religions (including various mainstream branches of Christianity), but they are flatly incompatible with scenarios typically portrayed by proponents of Creation Science and Intelligent Design.

Humans as Intelligent Designers

That any logistic hurdles to a supernatural engineer might be insuperable is further denied by the fact that even human beings—we mere mortals—now have the technological capacity to modify and sometimes reformat human biochemistry and genomic operations in ways that seem to improve upon what nature alone has accomplished. Indeed, the fields of medicine and pharmacology rest heavily on the premise that human health can be improved by biochemical interventions, such as when insulin is administered to patients with diabetes or when vitamin pills are swallowed to compensate for the body's inability to manufacture ascorbic acid. If humans can intervene in such ways, should any less be expected of an omnipotent and loving supernatural designer? One standard theological response—not necessarily incompatible with evolutionary biology but incompatible with standard Creation Science—is that good deeds by people with free will evidence the beneficence of a supreme God who ultimately made all of this possible, by setting in motion natural processes that eventuated in the evolutionary emergence of an intelligent biological species (*Homo sapiens*) that is now itself capable of performing "medical miracles" and other highly admirable technological feats.

The therapeutic macromolecules of pharmacology traditionally were extracted as medicinal compounds from various plants and animals in nature. In recent decades, scientists have gone

one big step further by artificially evolving novel proteins in the laboratory.[16] The process begins when biochemists extract or synthesize proteins that then are subjected to mutational changes in a test tube. From out of this process come countless protein variants that can be artificially selected for particular biochemical properties, such as an enhanced catalytic effectiveness on a specified (and sometimes novel) molecular substrate. New pharmaceuticals have been invented by this artificial procedure of "directed evolution" in the laboratory. The director in this case is neither some supernatural agent nor nature per se, but rather mere mortal biochemists who have learned to apply some of the scientific lessons from natural selection and evolutionary biology. Again, some theologians might interpret all of this as evidence of God's great foresight and beneficence in setting up the original conditions for life and human evolution, such that human intelligence and all of its resulting products are themselves prima facie evidence of God's handiwork. Be that as it may, directed protein evolution indicates that improved biological designs in macromolecules, beyond what nature (or God) alone had accomplished directly, are entirely plausible. Demonstrably, even we humans can engineer such betterments.

Therapeutic drugs conventionally were delivered by injection or ingestion, but in recent years human genetic engineers have gained even greater technological capabilities—in this case, to alter human genes directly and thereby coax our tissues to produce desirable biological compounds endogenously. I'm referring to the diverse procedures of genetic engineering and gene therapy wherein scientists ultra-microsurgically cut and paste particular pieces of DNA from one genome to another, using a variety of recombinant DNA technologies. The piece of DNA that is artificially transferred into a recipient's body is called a transgene. It may have originated from another species or it may have come from another member of the same species.

A promising recent example of real-life human gene therapy involved severe combined immunodeficiency disease (SCID), also known as "bubble boy disease" because children with this immunological disorder sometimes must be isolated, in a bubble, to stop possible infections. SCID often arises from defects in a gene for the enzyme adenosine deaminase (ADA), which otherwise breaks down compounds that impair key components of the human immune system. Early in 2009, Dr. Alessandro Aiuti and his research group reported results of a scientific study in which nondefective copies of an ADA gene were transferred (using a retroviral vector) into each of 10 young children with SCID.[17] Normally, SCID is a fatal disorder as the body is left defenseless against infections, but all 10 children who received the transgene survived the trial periods (which had a median of four years); and, importantly, their immune systems appeared to have been substantially restored by the gene therapy.

Although gene therapy as a medical discipline is still in its infancy, its procedures hold great promise. It is hard to dispute, for example, that replacing a defective gene that produces SCID, or other terrible genetic disorders such as Huntington disease, with an operative version that restores a person's health would represent a genuine improvement in the molecular design of the patient's genome. The technological capacity to isolate and insert transgenes into patients is already available, and it will be increasingly deployed by the medical profession in the coming years. The rapidly emerging field of genetic engineering—of which human gene therapy is just one small part—is much too vast to address here; interested readers are directed elsewhere for a broad-sweeping discussion of this fascinating topic.[18]

Suffice it here to state the following: to ascribe flaws in the human genome to insuperable technical hurdles for the Intelligent Designer could be taken to imply that a benevolent Creator God was even more challenged, technologically, than are we mere humans that He supposedly crafted in His image.[19] Thus, the Leibnizian notion that an omnipotent and loving God directly engineered the best of all possible biotic worlds, in this case with respect to the human genome, is patently incorrect.

NATURAL THEOLOGY AND THEODICY REVISITED

The many inevitable imperfections in the human genome raise profound empirical and philosophical conundrums for Intelligent Design and other guises of Creation Science. In effect, these molecular flaws extend the centuries-old theodicy dilemma—previously evident in the arena of human morphological phenotypes and behaviors—into biochemistry's submicroscopic realm. Indeed, no finer level of molecular dissection (e.g., at the strata of atoms or subatomic particles) can overturn the scientific conclusion that pervasive defects of design extend well into even the most elemental physical arenas of human existence.

Of course, not everyone will agree with this interpretation. Michael Behe emphasized the exact converse when he focused on the (many indisputable) biochemical elegances of eukaryotic cells. These led Behe to conclude (in *Darwin's Black Box*, pp. 232–233):

> The result of these cumulative efforts to investigate the cell—to investigate life at the molecular level—is a loud, clear, piercing cry of "design!" The result is so unambiguous and so significant that it must be ranked as one of the greatest achievements in the history of science... The observation of the intelligent design of life is as momentous as the observation that the earth goes around the sun or that disease is caused by bacteria or that radiation is emitted in quanta.

Unfortunately, such hyperbole is unfounded for several reasons. First, nearly all post-Darwin biologists have understood that insentient natural selection can forge biological beauty and complexity without direct intelligent intervention; and modern geneticists have published extensive evidence that evolutionary interpretations likewise apply to molecular and genomic features that otherwise might be deemed irreducibly complex. Second, even if there was some merit to Behe's general sentiment that nature's beauty and complexity are prima facie evidence for intelligent craftsmanship, this type of "argument from design" is hardly novel. For example, Reverend William Paley eloquently elaborated the idea in *Natural Theology*, published in 1802; and, as emphasized by David Sedley in *Creationism and Its Critics in Antiquity*,[20] the argument from design traces back at least to the classical Greek philosopher Socrates more than 400 B.C.

Third, the argument from design neglects a powerful counterargument to design that likewise is age-old. In 1779 (about 20 years before the appearance of William Paley's *Natural Theology*), the Scottish philosopher-historian David Hume pithily captured both the argument from design and the counterargument to design in a verbal exchange between Cleanthes and Philo (two fictional characters in *Dialogues Concerning Natural Religion*):

> CLEANTHES: The Author of Nature is somewhat similar to the mind of man, though possessed of much larger faculties, proportioned to the grandeur of the work he has executed... By this argument alone, do we prove at once the existence of a Deity.
>
> PHILO: What surprise must we entertain, when we find him a stupid mechanic.

Like Hume, Charles Darwin himself was well aware that biological imperfection provided a powerful counterargument to intelligent design. In chapter 14 of *The Origin*, Darwin writes, "on the view of each organic being and each separate organ having been specially created, how utterly inexplicable it is that parts... should so frequently bear the plain stamp of inutility." Much more recently, a well-respected evolutionary biologist and philosopher summed up the situation thusly: "Paley's argument suffers from a fatal flaw, namely, the pervasiveness of deficiencies, dysfunctions, oddities, and cruelties in organisms."[21]

From the vast scientific evidence—genetic and biochemical—for biological design flaws (the subject of the current treatise), we now know that the "argument from imperfection" also applies with full force to submicroscopic molecular features in human DNA. And, from our modern understanding of evolutionary-genetic processes, it seems clear that the many abject failures (as well as the apparent successes) of molecular design, including pervasive flaws inside the human genome, have an immediate or direct etiology in unconscious natural processes rather than either ethereal processes or intelligent design.

Michael Behe is quite aware that imperfections exist at the biochemical and molecular levels, so in response to the argument from imperfection he writes the following (in *Darwin's Black Box*, p. 223):

> The most basic problem is that the argument (from imperfection) demands perfection at all. Clearly, designers who have the ability to make better designs do not necessarily do so. For example, in manufacturing, "built-in obsolescence" is not uncommon—a product is intentionally made so it will not last as long as it might, for reasons that supersede the simple goal of engineering excellence. Another example is a personal one: I do not give my children the best, fanciest toys because I don't want to spoil them, and because I want them to learn the value of a dollar. The argument from imperfection overlooks the possibility that the designer might have multiple motives, with engineering excellence oftentimes relegated to a secondary role...Another problem with the argument from imperfection is that it critically depends on psychoanalysis of the unidentified designer. Yet the reasons that a designer would or would not do anything are virtually impossible to know unless the designer tells you specifically what those reasons are.

Ironically, Behe's dismissal of the argument from imperfection in effect demolishes Intelligent Design as a testable scientific hypothesis. Under Behe's formulation above, ID is impervious to logic and empirical evidence because we can never know the mind or motives of the designer. But if we cannot draw objective inferences about the designer from the many well-documented flaws of biological craftsmanship, then neither can we make logical inferences about the creator from any suspected artistries of design. So, Behe cannot have it both ways; he cannot promote ID as a valid scientific argument in one breath while effectively annihilating ID as a scientific hypothesis in another.

When describing the mechanistic agencies of evolution (including natural selection), I prefer the term "non-intelligent" (over unintelligent or stupid) because the word more accurately conveys a nonjudgmental view of how differential survival and reproduction translate into adaptive evolution whenever genetic variation in fitness exists. Natural selection, genetic drift, mutation, and other evolutionary forces are neither stupid nor smart. They are nonsentient processes, as mindless as they are inevitable. Little wonder, then, that their biological outcomes often fall far short of designer perfection. The scientific interpretation is that inborn errors of metabolism and the many other basic architectural and operational flaws of the human genome are unavoidable evolutionary by-products of insentient natural forces.

If, on the other hand, natural causation is denied, and a caring Intelligent Designer is to be held directly responsible for life's imperfect features, then the theodicy challenge remains poignant. How could a Creator God have engineered such a deeply flawed biological world, right down to its most elemental molecular features? Unless we pretend that biological defects do not exist, we seem forced to conclude that any Intelligent Designer is either technically fallible, morally challenged, or both.[22] Furthermore, if the intelligent designer is deemed to be the Abrahamic God (rather than a Martian, for example), then are we not guilty of blasphemy in ascribing to Him a direct hand in sponsoring the molecular genomic flaws that plague human health?

For some theologically conservative Christian denominations, one ready "escape" from this philosophical conundrum is to posit that imperfection in living systems results from the Fall from Grace in the Garden of Eden, and that humanity is now in a period of "devolution" through natural processes. Thus, perfection was the initial biological state from which imperfection descended. Under an evolutionary interpretation, by contrast, imperfection in living systems has always been present, and will remain so as humanity continues to evolve via nonsentient natural forces. Although biological structures and functions may often improve across the generations under the influence of natural selection, absolute perfection in the biological world apparently has been unachievable for all of the evolutionary reasons discussed in this book.

We now understand that the human genome is permeated with molecular imperfections ranging from the subtle to the egregious. Exactly how a Fall from Grace in the Garden of Eden might have become translated into these molecular defects is mechanistically unclear (to say the least). By contrast, how such genomic flaws arise and persist poses no insuperable mystery from the scientific perspective of evolutionary genetics.

Furthermore, molecular imperfections in the human genome provide significant evidence for evolution not only because they are imperfect but also because they are phylogenetically interpretable. Most genomic flaws (apart from de novo mutations that are currently confined to particular individuals) are distributed across biological taxa in ways that make evolutionary (i.e., phylogenetic) sense. This is true at all levels in the phylogenetic hierarchy. At the microevolutionary scale, many genetic disorders in humans "run in families" according to specifiable rules of Mendelian inheritance. And at the mesoevolutionary and macroevolutionary scales, humans share many molecular features, including particular molecular flaws, with various other taxa in the nested hierarchies of phylogeny. When fine details of molecular errors recur in phylogenetically related species, special-creation explanations for such errors are thus effectively eliminated (unless we are to suppose that a bumbling Creator made the same molecular mistakes time and again when directly forging different species).

EMANCIPATION AND RECONCILIATION

Evolution by natural causes emancipates religion from the shackles of theodicy. No longer need we agonize about why a Creator

God is the world's leading abortionist and mass murderer. No longer need we query a Creator God's motives for debilitating countless innocents with horrific genetic conditions. No longer must we anguish about the interventionist motives of a supreme intelligence that permits gross evil and suffering in the world. No longer need we be tempted to blaspheme an omnipotent Deity by charging Him directly responsible for human frailties and physical shortcomings (including those that we now understand to be commonplace at molecular and biochemical levels). No longer need we blame a Creator God's direct hand for any of these disturbing empirical facts. Instead, we can put the blame squarely on the agency of insentient, natural evolutionary causation.

In part for this reason, the evolutionary biologist and philosopher Francisco J. Ayala has hailed the discovery of natural selection as "Darwin's gift to science *and* religion."[23] As phrased by Ayala (p. 159), there is a genuine

> irony that the theory of evolution, which at first had seemed to remove the need for God in the world, now has convincingly removed the need to explain the world's imperfections as failed outcomes of God's design. Indeed, a major burden was removed from the shoulders of believers when convincing evidence was advanced that the design of organisms need not be attributed to the immediate agency of the Creator, but rather is an outcome of natural processes. If we claim that organisms and their parts have been specifically designed by God, we have to account for the incompetent design of the human jaw, the narrowness of the birth canal, and our poorly designed backbone [and, as this book has emphasized, multitudinous ill-conceived molecular features of the human genome as well—JCA] . . . The God of love and mercy could not have planned all this.

Ayala went on to write (p. 160),

> Proponents of ID would do well to acknowledge Darwin's revolution and accept natural selection as the process that accounts for the design of organisms, as well as for the dysfunctions, oddities, cruelties, and sadism that pervade the world of life. Attributing these to specific agency by the Creator amounts to blasphemy.

> Proponents and followers of ID are surely well-meaning people who do not intend such blasphemy, but this is how matters appear to a biologist concerned that God not be slandered with the imputation of incompetent design.

So, from this refreshing perspective, evolution can and should be viewed as a form of philosophical salvation (rather than as the inherent philosophical nemesis) of theology and religion.

Creation Science and the Intelligent Design movement are outgrowths of a particular idiosyncratic brand of Christian fundamentalism in America, and, thus, traditionally have been interpreted to fall within the sphere of religion as opposed to science (figure 5.2, top). However, religion writ large—including mainstream Christianity and most other nonfundamentalist faiths—can be compatible with the hard sciences, including molecular

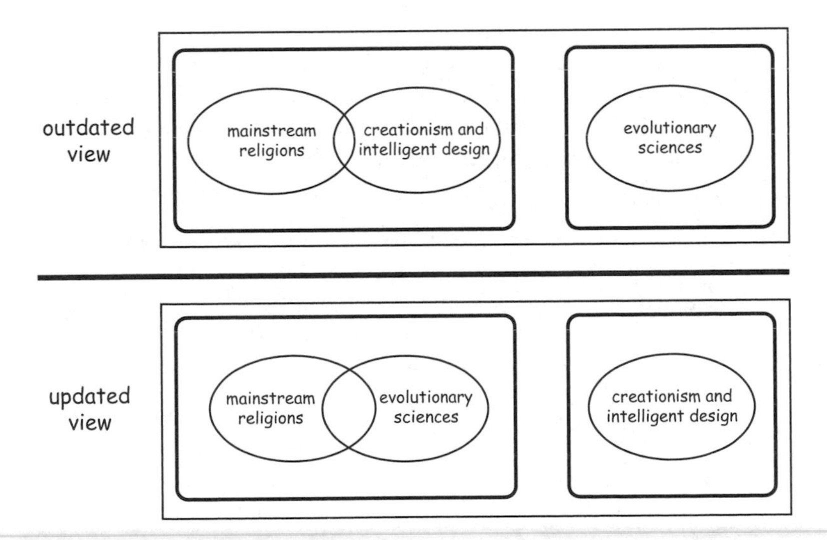

FIGURE 5.2. *Top*, the traditional placement of evolutionary biology as the odd man out to the spheres of mainstream religion and Intelligent Design in philosophical discourses about the human condition; *bottom*, an enlightened perspective in which Intelligent Design is the odd man out to the overlapping spheres of religion and the evolutionary sciences.

genetics and evolutionary biology. In recent years especially, many leading scientists, philosophers, and theologians have stressed that open-minded religion and evolutionary science need not be in opposition but instead can be viewed as interactive spheres in a joint endeavor to understand the context and meaning of human existence (figure 5.2, bottom).

This welcome sentiment—that the evolutionary sciences and religion both have important and complementary roles to play in philosophical discussions about the human condition—has been expressed in many notable statements, such as,[24]

> There is no contradiction between an evolutionary theory of human origins and the doctrine of God as Creator.—General Assembly of the Presbyterian Church

> ...new findings lead us toward the recognition of evolution as more than an hypothesis. In fact it is remarkable that this theory has had progressively greater influence on the spirit of researchers, following a series of discoveries in different scholarly disciplines. The convergence in the results of these independent studies—which was neither planned nor sought—constitutes in itself a significant argument in favor of the theory.—Pope John Paul II[25]

> ...science and religion are different ways of understanding. Needlessly placing them in opposition reduces the potential of both to contribute to a better future.—*Science, Evolution, and Creationism,* National Academy of Sciences and the Institute of Medicine, National Academies Press, Washington, D.C., 2008

> We the undersigned, Christian clergy from many different traditions, believe that the timeless truths of the Bible and the discoveries of modern science may comfortably coexist. We believe that the theory of evolution is a foundational scientific truth, one that has stood up to rigorous scrutiny and upon which much of human knowledge and achievement rests. To reject this truth or to treat it as "one theory among others" is to deliberately embrace scientific ignorance and transmit such ignorance to our children. We believe that among God's good gifts are human minds capable of critical thought and that the failure to fully employ this gift is a rejection of the will of the

> Creator...We urge school board members to preserve the integrity of the science curriculum by affirming the teaching of the theory of evolution as a core component of human knowledge. We ask that science remain science and that religion remain religion, two very different, but complementary, forms of truth.—*The Clergy Letters Project* signed by >10,000 Christian clergy members.

> In my view, there is no conflict in being a rigorous scientist and a person who believes in a God who takes a personal interest in each one of us. Science's domain is to explore nature. God's domain is the spiritual world, a realm not possible to explore with the tools and language of science. It must be examined with the heart, the mind, and the soul.—Francis Collins, director of the Human Genome Project

Ironically, the human genome's many inherent flaws—uncovered by painstaking genetic research and interpreted in the light of the evolutionary sciences—may be a glorious blessing in disguise for theology and religion. Structural and functional defects of genomic design—at the cell's most basic molecular level—highlight once again the futility of theodicic rationalizations for the human condition. They also point strongly toward evolutionary interpretations as opposed to direct intelligent design for life. From this vantage point, the evolutionary sciences can rightly be viewed as a welcome philosophical partner, rather than as an inherent antagonist, of nonfundamentalist religions. No longer need theologians be apologists for God with regard to the details of biology.

The evolutionary-genetic sciences thus can help religion to escape from the profound conundrums of Intelligent Design and thereby return religion to its rightful realm—not as the secular interpreter of the biological minutiae of our physical existence but rather as a respectable philosophical counselor on grander matters including ethics and morality, the soul, spiritual-ness, sacredness, and other such matters that have always been of ultimate concern to humanity.

Epilogue

Science is many things, but most of all it is mankind's surest and probably only reliable path to an *unprejudiced* understanding of natural phenomena. A standard dictionary definition of science—"knowledge ascertained by observation and experiment, critically tested, systematized and brought under general principles"[1]—captures the essence but neither the full richness nor the panorama of this remarkably successful approach to rational inquiry. Imagine a world without science, or even without the objective scientific discoveries of just the last five centuries (where most science has been concentrated). We might still believe that the Earth is the center of the universe, that life as well as the heavens arose de novo a few thousand years ago, that matter is composed of phlogiston, and that diseases are caused by bad night air. In short, we might believe almost anything, and we would have no way (other than by revelation or appeals to extra-natural causation) to decide among competing possibilities. Science provides a means of escape from this otherwise hopeless intellectual cul-de-sac.

Science expands the senses as well as the mind. Sensitive instruments of many sorts have uncovered entire spectrums of electromagnetic waves beyond our vision, sound frequencies beyond our hearing, pheromones and other scents beyond our olfactory capacity, and natural stimuli outside our physical senses of touch. Telescopes have allowed us to look outward in space and thereby also far back into the depths of time. The microscope introduced us to the remarkable world of bacteria and other minuscule forms of life that are far too small to see with the unaided eye. Biochemical tools of molecular biology have permitted us to peer even deeper into the fundamental makeup and internal operations of cells and genomes.

Science continually challenges us to expand our "common senses" as well. From mathematical calculations based on careful observations of celestial movements, Copernicus was able to deduce (counter to intuition or all prima facie appearances) that the Earth is not the center of the cosmos but instead is merely one of several small sun-revolving planets in one of probably untold numbers of solar systems. Likewise, from detailed physical observations and mathematical logic, Einstein showed, beyond all intuition to the contrary, that space and time are interrelated, and that matter and energy are interconvertible (according to a deceptively simple equation involving the square of the speed of light).

In biology, several scientists in the late 18th and early 19th centuries (among them Jean-Baptiste Lamarck, Comte de Buffon, and Erasmus Darwin) were well aware that forms of life mutate and change, but not until Charles Darwin (and his contemporary, Alfred Russel Wallace) was natural selection finally illuminated as the key directive force in biological evolution. Judging by the oft-negative reactions of Western societies to Darwinism despite overwhelming scientific evidence in its favor, natural selection has arguably been the least palatable of all scientific challenges to "common sense." Reverend William Paley and legions of other natural theologians, including their modern counterparts (proponents of Creation Science and Intelligent Design), have been either unable or unwilling to accept the demonstrable fact that biological outcomes in nature can arise from natural nonsentient processes rather than requiring direct and immediate intervention by either intelligent or supernatural forces. The immunity of Creation Science and its intellectual offshoot—ID—to evidence-based revision identifies these social movements as doctrinal rather than testable and thereby disqualifies them as science. It should likewise disbar them from serious consideration in *science* classrooms (except insofar as these philosophical crusades are interesting and important sociopolitical phenomena).

Is evolution itself subject to critical scientific evaluation and possible revision? Yes, absolutely. Indeed, if for no other reasons than scientific respect and personal acclaim, any serious research biologist would be overjoyed rather than dismayed to uncover convincing empirical evidence that appreciably modified or even overthrew any aspect of contemporary Darwinian or Mendelian thought, for example. He or she would proudly document and announce the findings, preferably in leading scientific journals, and a scientific audience would greet the news with excitement but skepticism. Soon other researchers would begin devising further experimental or observational tests of the phenomenon in question, hoping to be among the first to either support or refute the original claims. The new phenomenon would eventually stand or fall on its empirical and logical merits rather than on doctrinal beliefs.

In recent years, this revisionary scientific process has occurred time and again, for example, in the molecular branch of evolutionary genetics. As we have seen, the discoveries of introns, cytoplasmic genomes, pseudogenes, transposable and other repetitive elements, micro-RNAs, and various other classes of genetic material have all prompted major revisions of thought not only about genomic structure but also about fundamental evolutionary processes including the modes and levels of action of natural selection. It turns out that these discoveries (to date) have mostly supported rather than refuted Darwinian and Mendelian reasoning, but in principle this need not have been the case (if nature operated differently). Science cares only that the objective truth be uncovered, wherever it may lie. And all "truths" in science remain provisional, pending further critical evaluation.

It is entirely possible that some aspects of nature will yet be uncovered that completely defy scientific explanation based on any currently known natural processes. Must the default hypothesis of extra-naturalism then be accepted? No, because this would only stifle further scientific investigation into the matter. Imagine if science had been terminated forever by the mysticisms of the

Middle Ages, or perhaps by religious injunctions at any prior times. Clearly, we would be left today with a profound ignorance about mechanisms underlying physical and biological phenomena. We would also be missing all of the benefits (and problems) created by science-based technologies and medicine. Although some of us might wish at times to return to simpler days when we could suppose that all happenings in our world are under a loving God's blessed direct control, it is hard to imagine genuinely embracing a complete ignorance of natural causes once hard-won scientific knowledge had been gained.

The intent of this book has been to explore the insides of the human genome, one of the last bastions for possible biological evidence of intelligent design. Although not all of the molecular findings are in, and although scientific explanations always remain provisional (by definition), current indications are that the human genome, like previously studied macroscopic features of life, is a sophisticated yet highly tinkered and flawed contrivance of natural selection and other unconcerned evolutionary genetic mechanisms. To encapsulate such findings about the human genome, I would like to close this book by adapting Charles Darwin's famous concluding paragraph in *The Origin of Species* to the new molecular realities:

> It is interesting to contemplate a tangled mass of cells, each housing subcellular genetic compartments, recombining nuclear chromosomes of many lengths, with various kinds of transposable and viral elements flitting about, and with edited and nonedited RNA sequences crawling about its damp interiors, and to reflect that these elaborately constructed forms, so different from each other, and dependent upon each other in so complex a manner, have all been produced by laws acting around us.... There is grandeur in this view of the genome, with its several powers, having perhaps been originally breathed by the Creator into one or a few primordial molecular forms; and that, whilst this planet goes cycling on according to the fixed law of gravity, from so simple a beginning most beautiful, sometimes most awful, but always wondrous genomic features have been, and are being evolved.

NOTES AND REFERENCES

Chapter 1. The Eternal Paradox

1. Lander, E. S. and 243 others, 2001, Initial sequencing and analysis of the human genome, *Nature* **409**, 860–921; Venter, J. C. and 273 others, 2001, The sequence of the human genome, *Science* **291**, 1304–1351. These papers presented preliminary or draft sequences of the human genome; a more finalized product appeared three years later from the International Human Genome Sequencing Consortium: Finishing the euchromatic sequence of the human genome, *Nature* **431**, 931–945.

2. The statement has been attributed to Jesus (Matthew 27:46, King James Version of the Bible) shortly before his crucifixion.

3. A 1981 book entitled *When Bad Things Happen to Good People*, by Rabbi Harold S. Kushner, became a national best seller in the United States. It took the arguably comforting stance that many things in life happen not because God necessarily wants them to but rather for no special reason beyond randomness and happenstance.

4. For extended philosophical discussions on the problem of evil from various deductive and inductive arguments, see the following books: Hick, J., 2007, *Evil and the God of Love*, 2nd edition, Macmillan Press, London; Reichenbach, B. R., 1982, *Evil and a Good God*, Fordham University Press, New York.

5. Both quotations are from p. 154 in Sober, E., 2008, *Evidence and Evolution: The Logic Behind the Science*, Cambridge University Press, Cambridge, UK.

6. Dobzkansky, T., 1973, Nothing in biology makes sense except in the light of evolution, *Amer. Biol. Teacher* **35**, 125–129.

7. For example, Einstein once stated (see *Science, Philosophy, and Religion, A Symposium*, held in New York City in 1941) that he believed in a god who "reveals himself in the orderly harmony of what exists, not in

a god who concerns himself with the fates and actions of human beings." Near the end of his career, Einstein further declared: "In their struggle for the ethical good, teachers of religion must have the stature to give up the doctrine of a personal God; that is, give up the source of fear and hope which in the past placed such vast power in the hands of priests." The theoretician Paul Davies takes a similar stance when he argues (in *God and the New Physics*, 1983, Simon & Schuster, New York) that physics offers a surer path to God than does religion. Other leading physicists have held varying views on the existence of a god. For example, Isaac Newton believed that the Scriptures and the book of nature were equally real and that God made both available for people to study.

8. Subjects such as aging and death, sex and reproduction, and possible origins and meaning in life. I have borrowed the phrase from *The Biology of Ultimate Concern* (1967; New American Library, New York), a personal treatise on such matters by the evolutionary biologist Theodosius Dobzhansky.

9. Most modern historians have discontinued the use of "Dark Ages." They prefer instead less judgmental titles such as Late Antiquity or the Early Middle Ages.

10. A firm belief in God is what motivated many Christian natural theologians to explore and interpret nature's (i.e., God's) design. Similarly, because the Koran stresses that God's messages may be delivered through the natural world, some Muslim scholars have sought to examine their surroundings with "scientific" curiosity and attentiveness.

11. Copernicus's full treatise, *De Revolutionibus Orbium Coelestium* ("the revolution of celestial spheres"), was published in 1543, the year of his death. It came with an unauthorized preface (by Osiander) that attempted to placate the Church's criticisms of the heliocentric view. An even more famous clash between Copernican and religious thought came about a century later when, in 1633, a Pope-mandated inquisition found the Italian astronomer and scholar Galilei Galileo guilty of heresy for promoting (in *Dialogue Concerning the Two Chief Systems of the World: Ptolemaic and Copernican*) cosmological views that the Catholic Church deemed to contradict the Holy Scriptures.

12. This is not to imply that attempts at objective inquiry were nonexistent before Copernicus. In many other times and places, various people have sought to understand natural processes without undue recourse to occult or supernatural hypotheses.

13. The full title was *Natural Theology: or, Evidences of the Existence and Attributes of the Deity, Collected from the Appearances of Nature.*

14. When the Reverend Francis Henry, Earl of Bridgewater, died in 1829, his last will and testament directed the president of the Royal

Society of London to identify and distribute funds to persons selected to write books "On the Power, Wisdom, and Goodness of God, as Manifested in Creation." The result was the Bridgewater Treatises, a series of eight works (some published in multiple volumes) as follows: *The Adaptation of External Nature to the Moral and Intellectual Constitution of Man*, Thomas Chalmers, 1833; *Chemistry, Meteorology, and the Function of Digestion*, William Prout, 1834; *On the History, Habits, and Instincts of Animals*, William Kirby, 1835; *The Hand: Its Mechanisms and Vital Endowments as Evincing Design*, Charles Bell, 1837; *Geology and Mineralogy Considered with Reference to Natural Theology*, Dean Buckland, 1837; *On the Adaptation of External Nature to the Physical Condition of Man*, John Kidd, 1837; *On Astronomy and General Physics*, William Whewell, 1839; and *Animal and Vegetable Physiology Considered with Reference to Natural Theology*, Peter Mark Roget, 1840.

15. This and several other quotations in this section (unless otherwise noted) were gathered from Keith Thomson's *Before Darwin* (2005, Yale University Press), to which interested readers are also directed to find an excellent in-depth historical treatment of mankind's attempts to reconcile God and nature.

16. Aquinas's four other proofs of God involved an argument from motion (movement requires a mover), an argument from efficient cause (happenings require a director), an argument from being (existence requires a maker), and an argument from hierarchy (gradations, such as from good to evil, require a ranker).

17. From the diaries of William Bartram, 1791, as published in *Travels of William Bartram* (1955, Dover Publications, New York).

18. From John Muir's *My First Summer in the Sierra* (1916, Houghton Mifflin, Boston).

19. Leeuwenhoek quotation from A. Schierbeek (ed.), *Measuring the Invisible World: The Life and Works of Antoni van Leeuwenhoek F.R.S* (1959, Abelard-Schuman, London).

20. This is not to say that religion necessarily underlies what eventually emerged as modern science. Mendel (as well as Newton, Pasteur, and other scientists) believed in a Creator God for the general reasons that Catholics believe in God, and not just on the grounds of natural theology. Only after a scientific alternative was illuminated, in 1859, did the standard philosophy of natural theology begin to decline within the scientific community.

21. Apart from *Natural Theology*, Darwin also read Paley's 1794 book *A View of the Evidences of Christianity*. Two other authors that Darwin mentioned in his autobiography as having been influential in his early thinking were John Hershel (in *Preliminary Discourse on the Study of Natural*

Philosophy, which dealt with scientific methods and the nature of proof) and Alexander von Humboldt (in *Personal Narrative*, which described this naturalist's adventures in South America).

22. *The Blind Watchmaker* is the title of one of Dawkin's books, published in 1986 by Norton (New York).

23. G. J. Mendel, 1865, *Versuche ueber Pfanzenhybriden*, Verhandlungen des Naturforschenden Vereins (Bruenn) 4:3–47.

24. My brief account of the history of the ID movement in the United States is distilled from an extended treatment in *Evolution Vs. Creation* by Eugenie Scott (2004; Greenwood Press, Westport, CT.).

25. Published by Regnery Gateway, Washington, DC.

26. Published by the Free Press, New York.

27. To view imaginable mousetraps with fewer than five working parts, readers can visit the following Web sites: http://udel.edu/~mcdonald/oldmousetrap.html, or http://udel.edu/%7Emcdonald/mousetrap.html.

28. By Rowan and Littlefield publishers, Lanham, MD. *No Free Lunch* followed an earlier book by Dembski with a similar theme: *The Design Inference: Eliminating Chance through Small Probabilities* (Cambridge University Press, 1998).

29. All excerpts in this section are requoted from *Science, Evolution, and Creationism*, 2008, National Academy of Sciences and the Institute of Medicine, National Academies Press, Washington, DC, 2008.

30. In *Darwin's Black Box*, Behe states, "For the record, I have no reason to doubt that the universe is the billions of years old that the physicists say it is."

31. An English translation of the title of a 1997 book review by N. W. Blackstone: Argumentum ad Ignorantiam, *Quarterly Review of Biology* **72**, 445–447.

32. In its narrow-sense definition, recombination is the formation of new assortments of genes via the Mendelian processes of meiosis and fertilization that characterize nearly all sexually reproducing species. In any generation of sexual reproducers, recombination normally generates far more novel genetic variety for potential selective scrutiny than do de novo mutations alone (although mutations are the ultimate source of the genetic variation that recombination subsequently shuffles). Under a broader definition, genetic recombination can also refer to any other natural or artificial operations (e.g., human-mediated genetic engineering) by which genes become shuffled.

33. This statement applies to specific genetic variants presented for natural selection. For many species, however, *rates* of genetic recombination have been shown to vary in adaptively important ways. For example, the relative frequencies of selfing (self-fertilization) and outcrossing

(cross-fertilization) in hermaphroditic organisms can themselves evolve in response to natural selection via their influence on adaptively relevant levels and patterns of genetic variation. Another important point is that although de novo mutations in effect are random evolutionary happenings, their mean fitness consequences likely are weighted toward harmful rather than beneficial effects on organisms. This is because any well-honed organismal structure or function is more likely to be compromised than improved by random mutational tinkering. Thus, most newly arising mutations are quickly weeded out by natural selection, and the relatively few that are retained in populations are usually those that either are fitness-neutral or fitness-beneficial to the organisms that house them.

34. In principle, this need not have been the scientific outcome. For example, in the early 1800s Jean Baptiste Lamarck proposed a theory of heredity in which environmental conditions predisposed organisms to acquire genetic alterations specifically suiting them for those environments. If heredity had turned out to occur as envisioned by Lamarck (rather than as discovered by Mendel later in that century), then a mechanistic feedback between the environment and genes underlying adaptations could have meant that particular genetic changes arise precisely as needed within any species.

35. This phrase, from a 1970 book of that same title by Jacques Monod, has become a popular shorthand for describing the joint stochastic and deterministic aspects of evolution.

36. Genetic drift is a stochastic evolutionary process that involves generation-to-generation changes in a population's genetic composition due to chance sampling of gametes that lead to progeny. Via genetic drift, some alleles (alternative forms of a gene) may happenstantially increase and others decrease in frequency even if those alleles have no differential causal impact on genetic fitness or organismal adaptations. Genetic drift is normally most pronounced in small populations (for the same reason that any other such statistical "sampling errors" tend to be larger when sample sizes are small).

37. Several genetic mechanisms are known that can produce relatively large and sudden changes in organismal phenotype. For example, "homeobox" genes are specific DNA sequences that exert regulatory control over an individual's development, and particular mutations at key homeobox loci can have dramatic phenotypic effects (such as converting a fly's antennae into legs). Although most such mutations are undoubtedly strongly selected against in nature, they do evidence the possibility that rapid phenotypic alterations are at least possible in some traits.

38. This is not to say that evolutionary change in particular biological features is inevitable. For example, scientists are aware of many so-called living fossils (such as horseshoe crabs) in which the general body appearance has remained essentially unchanged for many tens of millions of years. There is no exigency that evolutionary change takes place if, for example, existing phenotypes are well adapted to a particular environment, or if relevant genetic variation for the phenotype in question does not happen to arise.

39. To pick a simple example, any pride of lions that suddenly evolved the capacity for flapping flight would likely gain an immediate adaptive advantage over its grounded counterparts, but the historical legacies of felid ancestry make this evolutionary trajectory vanishingly improbable. The restraining effects of countless other such genealogical legacies, though often less obvious, are no less real.

40. Palumbi, S. R., 2001, *The Evolution Explosion: How Humans Cause Rapid Evolutionary Change*, W.W. Norton, New York.

41. For a philosophical discussion of why evolutionism does not necessarily mean atheism, see Ruse, M., 2001. *Can a Darwinian be a Christian?* Cambridge University Press, Cambridge, MA.

42. An example of the exceptionally rapid DNA sequencing technologies currently available is in Bentley, D. R. and 194 others, 2008, Accurate whole human genome sequencing using reversible terminator chemistry, *Nature* **456**, 53–59.

43. That person was J. Craig Venter, one of the research leaders in the original genomic sequencing effort of 2001. The paper announcing this achievement was co-authored by Venter himself: Levy, S. and 30 others, 2007, The diploid genome sequence of an individual human, *PLoS Biology* **5**, e254. At the time of this writing (November, 2008), seven complete human genomes have been sequenced fully. Another of those genomes was from James Watson, co-discoverer of DNA's double-helical structure. Of course, partial DNA sequences are available from the genomes of many thousands more humans. As the speed of DNA sequencing skyrockets and the cost of DNA sequencing plummets (the cost of sequencing a full genome may soon fall to $10,000 or less), the near-term future will undoubtedly see the completion of many more whole-genome analyses.

44. Two such papers are as follows: Wheeler, D. A. and 26 others, 2008, The complete genome of an individual by massively parallel DNA sequencing, *Nature* **452**, 872–876; and Wang, J. and 76 others, 2008, The diploid genome sequence of an Asian individual, *Nature* **456**, 60–65.

45. Church, G. M., 2005, The Personal Genome Project, *Molec. Systems Biol.* **1**, 1–3.

46. Some for-profit companies already have begun offering customers genome-wide screening of SNPs (single nucleotide polymorphisms), ostensibly as an informational health service. These and other contemplated genome-screening initiatives are not without critics, for a host of reasons relating to technological and interpretive challenges of the data and possible abuses regarding patient consent and the sanctity of personal genetic information. See, for example, the following: Taylor, P., 2008, When consent gets in the way, *Nature* **456**, 32–33; and Reardon, J., R. Hindmarsh, H. Gottweis, U. Naue, and J. E. Lunshof, 2008, Misdirected precaution, *Nature* **456**, 34–35.

47. Among evolutionary biologists, the late Stephen Jay Gould was a vocal proponent of the idea that science and religion deal with separate arenas of inquiry, that is, that they have "nonoverlapping magisteria." See Gould, S. J., 1998, *Leonardo's Mountain of Clams and the Diet of Worms,* Three Rivers Press, New York.

48. For example, a common theological view that is entirely consistent with at least some forms of Christianity is that a supernatural God, in creating, inculcated a universe with natural processes from which life emerged. In other words, God in effect designed life indirectly, by setting the physical stage for biological evolution. This view seems to be held by many Christians and evolutionary scientists alike. Another somewhat related sentiment held by some theologians is that mutation, selection, and evolution are bona fide but that these processes (or perhaps some subset of them) are not entirely natural and instead take place continually (or perhaps intermittently) under God's direct hand. For these and deeper philosophical discussions about potential compatibilities between science and religion, see the following books: Michael Ruse, 2001, *Can a Darwinian be a Christian?* (Cambridge University Press, 2001); Cobb, J. B., Jr. (ed.), 2008, *Back to Darwin—A Richer Account of Evolution,* William B. Eerdmans, Grand Rapids, MI.

CHAPTER 2. FALLIBLE DESIGN: PROTEIN-CODING DNA SEQUENCES

1. In 1931, Garrod published an influential follow-up entitled *The Inborn Factors of Inherited Disease.*

2. In 1944, Avery, MacLeod, and McCarty (*J. Exptl. Med.* **79**, 137–158) showed experimentally that nucleic acids are the genetic or hereditary material of life. In 1953, Watson and Crick (*Nature* **171**, 736–738) described the basic molecular structure of double-helical DNA.

3. For Krebs's personal account of how the tricarboxylic acid cycle was deduced, see *Perspect. Biol. Med.* **14**, 154–170 (1970).

4. Duarte, N. C., S. A. Baker, N. Jamshidi, I. Thiele, M. Mo, T. Vo, R. Srivas, and B. O. Palsson, 2007, Global reconstruction of the human metabolic network based on genomic and bibliomic data, *Proc. Natl. Acad. Sci. USA* **104**, 1777–1782.

5. The experiments in 1944 by Avery, MacLeod, and McCarty (see footnote 2) demonstrated that DNA (and not protein) was the physical agent that can transform hereditable features in bacteria. In 1952, similar experiments by Hershey and Chase (*J. Genet. Physiol.* **36**, 39–56) showed that DNA must be the heritable component of a phage virus when it invades and replicates in a bacterial host (because the phage's protein coat fails to enter the bacterial cells during the infection process).

6. *Proc. Natl. Acad. Sci. USA* **27**, 499–506.

7. That information flowed in this direction became known as the "central dogma," but later discoveries were to show that the central dogma was overly dogmatic. For example, cells can also convert RNA→DNA in some cases, in a process known as reverse transcription. Nevertheless, nucleic acids clearly are the blueprints of heredity from which proteins are constructed.

8. The "one-gene/one-enzyme" slogan is actually an oversimplification in at least two regards: (1) a gene technically encodes a polypeptide (a string of amino acids), and two or more such polypeptides sometimes must assemble before a functional protein emerges; and (2) a gene's nucleotide sequence sometimes contributes to the construction of two or more different proteins (as will be elaborated in chapter 3). Nevertheless, the basic sentiment that particular genes encode particular corresponding proteins is fundamentally correct.

9. In elegant experiments by Heinrich Matthaei and Marshall Nirenberg in 1961, the first of the 64 codon assignments was deduced: the triplet UUU (uracil-uracil-uracil, where uracil is the mRNA ribonucleotide that is the biochemical analogue of thymine in DNA) was proved to specify the amino acid phenylalanine. By 1966, the entire chart of genetic codes had been revealed. In 1968, Nirenberg shared a Nobel Prize with Robert Holley and Gobind Khorana for their collective contributions to interpreting the genetic code and revealing its functional role in protein synthesis.

10. Because de novo mutations sometimes cause serious genetic disabilities in humans and other species, they understandably have a bad reputation. It is worth remembering, however, that mutations are also the ultimate source of genetic variety, including variation that is shuffled in each round of sexual reproduction into novel and sometimes adaptive genetic combinations. So, some mutations (albeit a

minority) are beneficial to their bearers or their descendants, and these tend to be favored by natural selection. If the mutational process were somehow to cease entirely, evolution and life itself would eventually expire.

11. Reviewed in La Du, B. N., 2001, Alkaptonuria, pp. 2109–2123 in *The Metabolic and Molecular Basis of Inherited Disease,* 8th edition, C. R. Scriver, A. L. Beaudet, W. S. Sly, and D. Valle (eds.), McGraw-Hill, New York.

12. When a deleterious mutation at an autosomal gene behaves as a recessive allele, the obvious implication is that even one copy of the alternative ("wild-type") allele in a heterozygous individual is sufficient to carry out the gene's normal cellular function. Conversely, when a deleterious mutation behaves as a dominant allele, the gene's function is disrupted to the extent that a heterozygous individual fully expresses the genetic disorder.

13. See the review in Luzzatto, L., A. Mehta, and T. Vulliamy, 2001, Glucose 6-phosphate dehydrogenase deficiency, pp. 4517–4554 in *The Metabolic and Molecular Basis of Inherited Disease,* 8th edition, C. R. Scriver, A. L. Beaudet, W. S. Sly, and D. Valle (eds.), McGraw-Hill, New York.

14. See the review in Weatherall, D. J., J. B. Clegg, D. R. Higgs, and W. G. Wood, 2001, The hemoglobinopathies, pp. 4571–4636 in *The Metabolic and Molecular Basis of Inherited Disease,* 8th edition, C. R. Scriver, A. L. Beaudet, W. S. Sly, and D. Valle (eds.), McGraw-Hill, New York.

15. See the review in Welsh, M. J., B. W. Ramsey, F. Accurso, and G. R. Cutting, 2001, Cystic fibrosis, pp. 5121–5188 in *The Metabolic and Molecular Basis of Inherited Disease,* 8th edition, C. R. Scriver, A. L. Beaudet, W. S. Sly, and D. Valle (eds.), McGraw-Hill, New York.

16. Stanbury, J. B., J. B.Wyngaarden, and D. S. Fredrickson (eds.), 1960, *The Metabolic Basis of Inherited Disease,* McGraw-Hill, New York.

17. McKusick. V. A. (ed.), 1998, *Mendelian Inheritance in Man* (12th edition), Johns Hopkins University Press, Baltimore, MD.

18. Stenson, P. D., E. V. Ball, M. Mort, A. D. Phillips, J. A. Shiel, N. S. Thomas, S. Abeysinghe, M. Krawczak, and D. N. Cooper, 2003. The human gene mutation database (HGMD®): 2003 Update, *Human Mutat.* **21**, 577–581.

19. To help pinpoint particular genes underlying complex metabolic disorders, genome-wide association studies are often conducted wherein large numbers of people are screened for polymorphic markers at many genetic loci. A statistical association of particular markers with particular disorders suggests that a marked region of the genome might contribute to the disorder. For strengths and weaknesses of such association studies, see Bourgain, C., E. Génin, N. Cox, and F. Clerget-Darpoux, 2007, Are

genome-wide association studies all that we need to dissect the genetic component of complex human diseases? *European J. Human Genet.* **15**, 260–263. For other recent examples of serious attempts to dissect a complex medical condition—clinical obesity—see Emilsson, V. and 34 others, 2008, Genetics of gene expression and its effect on disease, *Nature* **452**, 423–428; and Chen, Y. and 21 others, 2008, Variations in DNA elucidate molecular networks that cause disease, *Nature* **452**, 429–435.

20. Ley, T. J. and 47 others, 2008, DNA sequencing of a cytogenetically normal acute myeloid leukaemia genome, *Nature* **456**, 66–72.

21. For a thorough description of GWASs and other approaches to genetic mapping of complex human genetic disorders, see Altshuler, D., M. J. Daly, and E. S. Lander, 2008, Genetic mapping in human disease, *Science* **322**, 881–888.

22. Cotton, R. G. H. and 28 others, 2008, The human variome project, *Science* **322**, 861–862.

23. This disorder and other bizarre genetic conditions are detailed in a book by Lisa Chiu (2006): *When a Gene Makes You Smell Like a Fish*, Oxford University Press, New York.

24. Source: *Britannica online Encyclopedia* (www.britannica.com/human genetic disease).

25. Wood, L. D. and 41 others, 2007, The genomic landscape of human breast and colorectal cancers, *Science* **318**, 1108–1113.

26. The topic of DNA repair can get very complicated. For a comprehensive overview of repair mechanisms in mammals including humans, see Sancar A., L. A. Lindsey-Boltz, K. Ünsal-Kaçmaz, and S. Linn, 2004. Molecular mechanisms of mammalian DNA repair and the DNA damage checkpoints, *Annu. Rev. Biochem.* **73**, 39–85.

27. A strikingly similar example involves glucose-6-phosphate dehydrogenase deficiency—humanity's most common known enzymopathy. Some genotypes again appear to confer enhanced malarial resistance, thereby generating selective pressures that help to maintain in high frequency some otherwise deleterious *G-6-PD* alleles in human populations (Tishkoff and 16 others, 2001, Haplotype diversity and linkage disequilibrium at human *G6PD*: recent origin of alleles that confer malarial resistance, *Science* **293**, 455–462).

28. A sentiment famously expressed Albert Einstein.

29. Slatkin M., 2004, A population-genetic test of founder effects and implications for Ashkenazi Jewish diseases, *Am. J. Human Genet.* **75**, 282–293; Risch N., H. Tang, H. Karzenstein, and J. Ekstein, 2003, Geographic distribution of disease mutations in Ashkenazi Jewish population supports genetic drift over selection. *Am. J. Human Genet.* **72**, 812–822.

CHAPTER 3. BAROQUE DESIGN: GRATUITOUS GENOMIC COMPLEXITY

1. This is not to imply that introns are fully dispensable once present in a genome. As discussed later in this chapter, introns can and do play some important functional and evolutionary roles in eukaryotic genomes. However, any operational benefits that introns now convey did not necessarily originate for those reasons; nor do the benefits necessarily imply that life could not have been designed better if an omnipotent force was directly involved.

2. Frischmeyer, P. A. and H.C. Dietz, 1999, Nonsense-mediated mRNA decay in health and disease, *Human Mol. Genet.*, **8**, 1893–1900; Philips, A. V. and T. A. Cooper, 2000, RNA processing and human disease, *Cell. Mol. Life Sci.* **57**, 235–249.

3. Already by 1992 (only 15 years after the discovery of split genes), more than 100 different examples of point mutations in the vicinity of splice junctions had been identified as the source of various genetic disabilities in humans (Krawczak, M., J. Reiss, and D.N. Cooper, 1992, The mutational spectrum of single base-pair substitutions in mRNA splice junctions of human genes: causes and consequences, *Human Genet.* **90**, 41–54). Such splice-site mutations typically compromise the accuracy or efficiency of mRNA splicing.

4. López-Bigas, N., B. Audit, C. Ouzounis, G. Parra, and R. Guigó, 2005, Are splicing mutations the most frequent cause of hereditary disease? *FEBS Letters* **579**, 1900–1903;

5. For arguments in favor of the introns-early view, see the following: Darnell, J. E., Jr., 1978, Implications of RNA-RNA splicing in evolution of eukaryotic cells, *Science*, **202**, 1257–1260; Doolittle, W. F., 1978, Genes in pieces: were they ever together?, *Nature* **272**, 581–582; and Roy, S. W., M. Nosaka, S. J. de Souza, and W. Gilbert, 1999, Centripetal modules and ancient introns, *Gene* **238**, 85–91. For arguments in favor of the introns-late view, see Cavalier-Smith, T., 1985, Selfish DNA and the origin of introns, *Nature* **315**, 283–284; Orgel, L. E. and F. H. Crick, 1980, Selfish DNA: the ultimate parasite, *Nature*, **284**, 604–607; and Palmer, J. D. and J. M. Logsdon, 1991, The recent origins of introns, *Curr. Opin. Genet. Develop.* **1**, 470–477.

6. Kidwell, M. G. and D. R. Lisch, 2000, Transposable elements and host genome evolution, *Trends Ecol. Evol.* **15**, 95–99; Lynch, M. and A. O. Richardson, 2002, The evolution of spliceosomal introns, *Curr. Opin. Genet. Dev.* **12**, 701–710; Rogers, J. H., 1990, The role of introns in evolution, *FEBS Letters* **268**, 339–343.

7. An intriguing observation is that a few introns (known as Group II introns) are "self-splicing" by virtue of being endowed with some of the

necessary molecular machinery necessary to effect their removal from pre-mRNAs. These are present in prokaryotes and sometimes in cytoplasmic genomes (but apparently not the nuclear genomes) of eukaryotes. Group II introns are evolutionarily intriguing because, unlike the more standard introns (known as spliceosomal introns), they in effect are mobile within the genome and thus bear some operational resemblance to genuine transposable elements. They are also evolutionarily intriguing because they might possibly have been involved somehow in the origin of the spliceosome.

8. Purugganan, M. and S. Wessler, 1992, The splicing of transposable elements and its role in intron evolution, *Genetica* **86**, 295–303.

9. Gilbert, W., 1978, Why genes in pieces? *Nature* **271**, 501.

10. Boue, S., I. Letunic, and P. Bork, 2003, Alternative splicing and evolution, *BioEssays* **25**, 1031–1034; Johnson, J. M. and nine others, 2003, Genome-wide survey of human alternative pre-mRNA splicing with exon junction microarrays, *Science* **302**, 2141–2144.

11. In other words, if the nucleotide space currently occupied by introns was converted entirely to exons, this subfraction of the human genome alone could house about 750,000 protein-coding genes, which would be far more than enough to satisfy all of the body's functional requirements.

12. Inside the core promoter of many genes (including about 32% of human genes) is a "TATA box": a short block of nucleotides (such as T-A-T-A-A-A) that specifically binds one of the transcription factors to which RNA polymerase in turn becomes bound. Additional transcription factors complete the assemblage; each molecular ensemble of promoter-bound RNA polymerase and transcription factors is termed a transcription initiation complex. Eukaryotic genes without a TATA box per se typically have other comparable transcription initiation sites. In the human genome, approximately 2,500 genes are devoted to encoding different transcription factors.

13. For a much more detailed account, see Tijan, R., 1995, Molecular machines that control genes, *Sci. Amer.* **272**, 54–61.

14. Culbertson, M. R., 1999, RNA surveillance: unforeseen consequences for gene expression, inherited genetic disorders and cancer, *Trends in Genet.* **15**, 74–70.

15. Manning, G., D. B. Whyte, R. Martinez, T. Hunter, and S. Sudarsanam, 2002, The protein kinase complement of the human genome, *Science* **298**, 1912–1934.

16. Four recent reviews are as follows: Selbach, M., B. Schwanhäusser, N. Thierfelder, Z. Fang, R. Khanin, and N. Rajewsky, 2008, Widespread changes in protein synthesis induced by microRNAs, *Nature* **455**, 58–63;

Baek, D., J. Villén, C. Shin, F. D. Camargo, S. P. Gygi, and D. P. Bartel, 2008, The impact of microRNAs on protein output, *Nature* **455**, 64–71; He, L. and G. J. Hannon, 2004, MicroRNAs: small RNAs with a big role in gene regulation, *Nature Rev. Genet.* **5**, 522–531; Alvarez-Garcia, I. and E. A. Miska, 2005, Micro-RNA functions in animal development and human disease, *Development* **132**, 4653–4662. In 2008, an issue of *Science* magazine (**319**, 1781–1799) also included several papers emphasizing known and suspected regulatory roles for miRNAs.

17. For more on long noncoding RNAs, see the following review: Petherick, A., 2008, The production line, *Nature* **454**, 1043–1045.

18. Misteli, T., 2007, Beyond the sequence: cellular organization of genome function, *Cell* **128**, 787–800.

19. Bernstein, B. E., A. Meissner, and E. S. Lander, 2007, The mammalian epigenome, *Cell* **128**, 669–681. Epigenetics refers to the entire suite of mechanisms, developmental pathways, and environmental influences on gene expression patterns.

20. Jones, P. A., and S. B. Baylin, 2002, The fundamental role of epigenetic events in cancer, *Nature Rev. Genet.* **3**, 415–428.

21. Weatherall, D. J., D. H. Higgs, W. G. Wood, and J. B. Clegg, 1984, Genetic disorders of human haemoglobin as models for analyzing gene regulation, *Phil. Trans. Royal Soc. London B*, **307**, 247–259.

22. Most eukaryotic genes have a poly-A tail (AAA...AAA) at their "downstream" end, past the gene's stop codon and trailer regions, that appears to facilitate the export of mRNA from the nucleus. Mutations in this tail can therefore have serious consequences for the gene's proper expression.

23. Badcock, C. and B. Crespi, 2008, Battle of the sexes may set the brain, *Nature* **454**, 1054–1055; for a broader review, see also the following and references therein: Badcock, C. and B. Crespi, 2006, Imbalanced genomic imprinting in brain development: an evolutionary basis for the aetiology of autism, *J. Evol. Biol.* **19**, 1007–1032.

24. Leask, J. A. Leask, and N. Silove, 2005, Evidence for autism in folklore? *Archives of Disease in Childhood* **90**, 271–272.

25. There are additional elements to the conflict theory of genetic imprinting, especially in mammal species with multisire broods. In such cases, a female's embryos are on average more closely related for maternally inherited genes than for paternally inherited genes. Thus, paternally expressed genes in an embryo are expected to be under selection pressure for selfishly extracting as many resources as feasible from the mother, whereas maternally expressed genes in an embryo should be under selection pressure for a more equitable compromise. The resulting selection pressures on genes have probably contributed to how

genomic imprinting has evolved to express or silence genes according to their parental origin and their effects on mother and offspring fitness. For much more on this broad topic, see the following: de la Casa-Esperon, E. and C. Sapienza, 2003, Natural selection and the evolution of genome imprinting, *Annu. Rev. Genet.* **37**, 349–370; Haig, D., 1993, Genetic conflicts in human pregnancy, *Q. Rev. Bio.* **68**, 495–532; Haig, D., 2000, The kinship theory of genomic imprinting, *Annu. Rev. Ecol. Syst.* **31**, 9–32; Wilkins, J.F., 2005, Genomic imprinting and methylation: epigenetic canalization and conflict, *Trends Genet.* **21**, 356–365.

26. Altmann, R, 1890, *Die Elementarorganismen und Ihre Beziehungen Zu Den Zellen*, Verlag von Veit, Leipzig, Germany, as cited in Graff, C., D. A. Clayton, and N.-G. Larsson, 1999, Mitochondrial medicine—recent advances, *J. Internal Med.* **246**, 11–23.

27. Two of the pioneering analyses were as follows: Brown, W. M., 1980, Polymorphism in mitochondrial DNA of humans as revealed by restriction endonuclease analysis, *Proc. Natl. Acad. Sci. USA* **77**, 3605–3609; and Cann, R. L., M. Stoneking, and A. C. Wilson, 1987, Mitochondrial DNA and human evolution, *Nature* **325**, 31–36. Since then, dozens of genealogical studies based on mitochondrial genes, often in conjunction with data from the nuclear genome, have revealed many historical demographic details about bygone human populations and their spatial movements on the planet (see the following and references therein: Avise, J. C., 2004, *Molecular Markers, Natural History, and Evolution*, 2nd edition, Sinauer, Sunderland, MA., pp. 298–301; Cela-Conde, C. J. and F. J. Ayala, 2007, *Human Evolution: Trails from the Past*, Oxford University Press, Oxford, UK).

28. Nass, M. M. K. and S. Nass, 1963, Intramitochondrial fibers with DNA characteristics, I, fixation and electron staining reactions, *J. Cell. Biol.* **19**, 593–611.

29. Anderson, S. and 13 others, Structure and organization of the human mitochondrial genome, *Nature* **290**, 457–465.

30. Technically, many individuals are actually heteroplasmic for mtDNA, meaning that two or more mtDNA sequences are jointly present. Nonetheless, one sequence usually greatly predominates, and any departures from that prototype are usually due to recent somatic mutations. The reasons for effective homoplasmy are not entirely clear, but one likelihood is that mtDNA molecules undergo severe bottlenecks (reductions in numbers) in germ-cell lineages (Chapman, R. W., J. C. Stephens, R. A. Lansman, and J. C. Avise, 1982, Models of mitochondrial DNA transmission genetics and evolution in higher eukryotes, *Genet. Res.* **40**, 41–57; Poulton, J., V. Macaulay, and D. R. Marchington, 1998, Mitochondrial genetics '98: is the bottleneck cracked?, *Am. J. Human*

Genet. **62**, 752–757). If so, this could account not only for the tendency for within-individual sequence homogeneity (since all mtDNAs in an individual's somatic cells trace back to the population of mtDNA molecules in the mother's oocyte) but also the rapid genetic transitions from one homoplasmic state to another that often have led to observed differences in mtDNA sequence among individuals in local populations.

31. The relatively high mutation rate in mtDNA also translates into a fast evolutionary pace for the molecule (as first documented by Brown, W. M., M. George, Jr., and A. C. Wilson, 1979, Rapid evolution of animal mitochondrial DNA, *Proc. Natl. Acad. Sci. USA* **76**, 1967–1971). Several factors probably contribute to this rapid mtDNA evolution: relatively inefficient mechanisms of DNA repair in mitochondria; exposure to the high concentrations of highly corrosive oxygen radicals within mitochondria; a relaxation of functional constraints due to the fact that mtDNA does not produce proteins directly involved in its own replication, transcription, or translation; and the fact that mtDNA is naked, that is, not tightly bound to histone proteins that are evolutionarily conservative and may generally constrain evolutionary rates in nuclear DNA.

32. Wallace, D.C. and others, 1988, Mitochondrial DNA mutation associated with Leber's hereditary optic neuropathy, *Science* **242**, 1427–1430. This study was headed by Doug Wallace, now at the University of California at Irvine, who has become a world leader in the field of human mitochondrial genetics.

33. Shoffner, J. M. and others, 2000, Myoclonic epilepsy and ragged-red fiber disease (MERRF) is associated with a mitochondrial DNA $tRNA^{Lys}$ mutation, *Cell* **61**, 931–937; Wallace, D. C. and others, 1988, Familial mitochondrial encephalomyopathy (MERRF): genetic, pathophysiological, and biochemical characterization of a mitochondrial DNA disease, *Cell* **55**, 601–610.

34. Copeland, W. C., J. T. Wachsman, F. M. Johnson, and J. S. Penta, 2002, Mitochondrial DNA alterations in cancer, *Cancer Invest.* **20**, 557–569; Horton, T. M. and seven others, Novel mitochondrial DNA deletion found in a renal cell carcinoma, *Genes Chromosomes Cancer* **15**, 95–101.

35. Included in the tallies are mitochondrial mutations per se, and single-gene nuclear mutations that seriously impair mitochondrial function. References are as follows: Chinnery, P. F. and 8 others, 2000, Epidemiology of pathogenic mitochondrial DNA mutations, *Ann. Neurol.* **48**, 188–193; Darin, N., A. Oldfors, A. R. Moslemi, E. Holme, and M. Tulinius, 2001, The incidence of mitochondrial encephalomyopathies in childhood: clinical features and morphological, biochemical, and DNA abnormalities, *Ann. Neurol.* **49**, 377–383; Schaefer, A. M., R. W. Taylor, D. M. Turnbull, and P. F. Chinnery, 2004, The epidemiology of

mitochondrial disorders—past, present, and future, *Biochem. Biophys. Acta* **1659**, 115–120; Skladal, J. Halliday, and D. R. Thorburn, 2003, Minimum birth prevalence of mitochondrial respiratory chain disorders in children, *Brain* **126**, 1905–1912.

36. Reasons for the former neglect of mitochondria by the medical community are several, but one involves mtDNA's peculiar mode of inheritance (maternal, and involving oft-large populations of molecules in gametes). Typically, medical geneticists were trained in the tradition of Mendelian inheritance, which applies to nuclear autosomal genes but not to the mitochondrial genome. Partly for this reason, diagnosticians often failed to identify the signatures of mitochondrial inheritance in various diseases presented by their patients.

37. McFarland, R., R. W. Taylor, and D. M. Turnbull, 2007, Mitochondrial disease—its impact, etiology, and pathology, *Curr. Topics Develop. Biol.* **77**, 113–155; Wallace, D. C., 2005, A mitochondrial paradigm of metabolic and degenerative diseases, aging, and cancer: a dawn for evolutionary medicine, *Annu. Rev. Genet.* **39**, 359–407.

38. Senescence, used here synonymously with aging, is defined as a persistent decline in the age-specific survival probability or reproductive output of an individual due to internal physiological deterioration.

39. A repository for such information is MITOMAP (http://www.MITOMAP.org), an organized database for human mtDNA (Brandon, M. C., M. T. Lott, K. C. Nguyen, S. Spolim, S. B. Navathe, P. Baldi, and D. C. Wallace, 2005, MITOMAP: a human mitochondrial genome database—2004 update, *Nucleic Acids Res.* **33**, D6121-D613).

40. Luft, R., D. Ikkos, G. Palmieri, L. Ernster, and B. Afzelius, 1962, A case of severe hypermetabolism of nonthyroid origin with a defect in the maintenance of mitochondrial respiratory control: a correlated clinical, biochemical, and morphological study, *J. Clinical Invest.* **41**, 1776–1804.

41. Margulis, L., 1981, *Symbiosis in Cell Evolution: Life and Its Environment in the Early Earth*, W. H. Freeman, San Francisco.

42. Why the process stopped before completion is not fully understood, but one possibility is that changes (drift) in the genetic code halted the process because mitochondrial genes could no longer be read properly by the ribosomes that translate nuclear DNA.

43. Gherman, A. and 9 others, 2007, Population bottlenecks as a potential major shaping force of human genome architecture, *PLOS Genetics* **3**, 1223–1231.

44. Hazkani-Covo, E., R. Sorek, and D. Graur, 2003, Evolutionary dynamics of large numts in the human genome: rarity of independent insertions and abundance of post-insertion duplications, *J. Molec. Evol.* **56**, 169–174.

45. Collura, R. V. and C.-B. Stewart, 1005, Insertions and deletions of mtDNA in the nuclear genomes of Old World monkeys and hominoids, *Nature* **378**, 485–489.

46. The disorder is Pallister-Hall syndrome, the symptoms of which include cleft palate, polydactyly, and brain malformations. A clinical case with numt involvement was reported by Turner, C. and eight others, 2003, Human genetic disease caused by de novo mitochondrial-nuclear DNA transfer, *Human Genet.* **112**, 303–309.

47. This observation, later confirmed and extended by many researchers, was first made in the late 1970s by Carl Woese and colleagues (Fox, G. E., L. J. Magrum, W. E. Balch, R. S. Wolfe, and C. R. Woese, 1977, Classification of methanogenic bacteria by 16S ribosomal RNA characterization, *Proc. Natl. Acad. Sci. USA* **74**, 4537–4541; Woese, C. R. and G. E. Fox, 1977, Phylogenetic structure of the prokaryotic domain: the primary kingdoms, *Proc. Natl. Acad. Sci. USA* **74**, 5088–5090).

CHAPTER 4. WASTEFUL DESIGN: REPETITIVE DNA ELEMENTS

1. Eichler, E. E., 2001, Recent duplication, domain accretion and the dynamic mutation of the human genome, *Trends in Genetics* **17**, 661–669; Bailey, J. A., Z. Gu, R. A. Clark, K. E. Rinert, R. V. Samonte, S. Schwartz, M. D. Adams, E. W. Myers, P. W. Li, and E. E. Eichler, 2002, Recent segmental duplications in the human genome, *Science* **297**, 1003–1007.

2. Zhang, Z. and M. Gerstein, 2004, Large-scale analysis of pseudogenes in the human genome, *Current Opin. Genet. Dev.* **14**, 328–335.

3. Lynch, M. and J. S. Conery, 2003a, The evolutionary demography of duplicate genes, *J. Struc. Func. Genomics* **3**, 35–44; Lynch, M. and J. S. Conery, 2003b, The origins of genome complexity, *Science* **302**, 1401–1404.

4. Li, W.-H., 1997, *Molecular Evolution*, Sinauer, Sunderland, MA; Campbell, N. A., J. B. Reece, and L. G. Mitchell, 1999, *Biology*, 5th edition, Benjamin-Cummings, Menlo Park, CA.

5. Weiner, A. M., P. L. Deininger, and A. Efstratiadis, 1986, Nonviral retroposons: genes, pseudogenes, and transposable elements generated by the reverse flow of genetic information, *Annu. Rev. Biochem.*, **55**, 631–666.

6. See the review in Balakirev and Ayala, 2003, Pseudogenes: are they "junk" or functional DNA?, *Annu. Rev. Genet.* **37**, 123–151. For example, a processed pseudogene might function by displaying a key regulatory site (such as a binding domain for transcription factors) that could influence the activities of adjacent genes along the chromosome. Tam and colleagues (2008) uncovered an empirical example (in mice)

of how pseudogene-derived RNA molecules can also play a role in gene regulation (Pseudogene-derived small interfering RNAs regulate gene expression in the mouse, *Nature*, **453**, 534–538). It is even conceivable that on rare occasions a useless pseudogene might spring back to life by fortuitously acquiring rehabilitating mutations.

7. C. Dib and others, 1996, A comprehensive genetic map of the human genome based on 5,264 microsatellites, *Nature* **380**, 152–154.

8. The total fraction of the human genome devoted to SSRs is thus greater than that devoted to protein-coding genes. A database for SSRs in *Homo sapiens* can be accessed at http://www/ccmb.res.in/ssr and http://www.ingenovis.com/ssr (Subramanian, S., V. M. Madgula, R. George, S. Kumar, M. W. Pandit, and L. Singh, 2003, SSRD: simple sequence repeats database of the human genome, *Comp. Func. Genomics* **4**, 342–345).

9. This sentiment does not preclude the possibility that some positive functions will yet be identified for SSRs. Because microsatellites are found in regulatory as well as coding regions of the genome, some of them are likely to have an evolved impact on how genes are expressed. For an example, see Albanese, V. and others, 2001, Quantitative effects on gene silencing by allelic variation at a tetranucleotide microsatellite, *Hum. Mol. Genet.* **10**, 1785–1792.

10. Strand slippage during DNA replication is the most likely mechanism underlying expansions and contractions at a microsatellite locus. In any long tract of tandem SSRs, the two complementary strands of DNA can easily misalign during replication, thereby leading to increases or deletions of sequence. Other mechanisms of SSR expansion and contraction are also known (see Sinden, S. R., 2001, Origins of instability, *Nature* **411**, 757–758).

11. Ashley, C. T., Jr., and S. T. Warren, 1995, Trinucleotide repeat expansion and human disease, *Annu. Rev. Genet.* **29**, 703–728; Cummings, C. J. and H. Y. Zoghbi, 2000, Fourteen and counting: unraveling trinucleotide repeat diseases, *Hum. Mol. Genet.* **9**, 909–916.

12. Esnault, C., J. Maestre, and T. Heidemann, 2000, Human LINE retrotransposons generate processed pseudogenes. *Nature Genet.* **24**, 363–367.

13. This process nonetheless can result in a proliferation of transposons within a genome, for the following reason. When a transposon excises from a chromosome, a double-strand break arises that often is repaired by a cell using the homologous chromosome as a template.

14. The process starts with standard transcription of a retrotransposon into RNA, followed by reverse-transcription of that RNA (catalyzed by the enzyme reverse transcriptase) into a new copy of DNA which then inserts at another location in the genome.

15. The following are merely a few among many publications supporting this point: Watkins, W. S. and 13 others, 2003, Genetic variation among world populations: inferences from 100 Alu insertion polymorphisms, *Genome Res.* **13**, 1607–1618; Vincent, B. J. and seven others, 2003, Following the LINEs: an analysis of primate genomic variation at human-specific LINE-1 insertion sites, *Molec. Biol. Evol.* **20**, 1338–1348; Nikaido, M. and ten others, 2001, Retroposon analysis of major cetacean lineages: The monophyly of toothed whales and the paraphyly of river dolphins, *Proc. Natl. Acad. Sci. USA* **98**, 7384–7389.

16. The *gag* gene encodes a capsid protein that helps sequester molecules necessary for reverse transcription, and the *pol* gene encodes protein domains (including reverse transcriptase per se) involved in mobile element movement. Compared to bona fide retroviruses, LTR elements lack only the *env* gene, which facilitates the entry of a retrovirus into a host cell. Whereas retroviruses can readily hop from one host organism to another, retroviral-like mobile elements usually confine their hopping to different sites within a host genome. Because of their similar features, many biologists think that viruses and retroviral-like transposable elements are evolutionarily interrelated (e.g., Xiong, Y. and T. J. Eichbush, 1990, Origin and evolution of retroelements based upon their reverse transcriptase sequences, *EMBO J.* **9**, 3353–3362).

17. D. A. Hickey, 1982, Selfish DNA, a sexually-transmitted nuclear parasite, *Genetics* **101**, 519–531.

18. This assumes that transmission of the mobile element is strictly "vertical," that is, from parent to offspring. For infectious elements (such as retroviruses) with horizontal transmission, sexual reproduction by the host is not typically a prerequisite for proliferation.

19. Indeed, Barbara McClintock—the original discoverer of mobile elements—termed jumping genes "controlling elements" (1956, Controlling elements and the gene, *Cold Spring Harbor Symp. Quant. Biol.* **21**:197–216) because of how they seemed to influence the expression of nearby functional genes. Several cases have since been documented in which the regulatory capabilities of mobile elements were evolutionarily recruited into host-beneficial services. For example, the expression in salivary glands of the amylase-1 gene (which codes for a starch-digesting enzyme) is due in part to regulatory influence by a retroviral-like element that inserted (about 45 million years ago) near the amylase locus. Other such examples involve two human globin genes that utilize truncated versions of formerly mobile *Alu* elements to regulate gene expression in a tissue-specific manner. For a recent example of how extensive genomic scans are helping to reveal such cases of regulatory influence by mobile elements, see Lowe, C. B., G. Bejerano, and D. Haussler, 2007,

Thousands of human mobile element fragments undergo strong purifying selection near developmental genes, *Proc. Natl. Acad. Sci. USA* **104**, 8005–8010.

20. Reviews by McDonald, F. J., 1993, Evolution and consequences of transposable elements, *Current Opin. Genet. Develop.* **3**, 855–864; McDonald, F. J., 1995, Transposable elements: possible catalysts of organismic evolution, *Trends Ecol. Evol.* **10**, 123–126; Brosius, J., 1999, RNAs from all categories generate retrosequences that may be exapted as novel genes or regulatory elements, *Gene* **238**, 115–134.

21. Brouha, B., J. Schustak, R. M. Badge, S. Lutz-Prigge, A. H. Farley, J. V. Moran, and H. H. Kazazian, Jr., 2003, Hot L1s account for the bulk of retrotransposition in the human population, *Proc. Natl. Acad. Sci. USA* **100**, 5280–5285; Cordaux, R., D. J. Hedges, S. W. Herke, and M. A. Batzer, 2006, Estimating the retrotransposition rate of human *Alu* elements, *Gene* **373**, 134–137.

22. Miné, M. and 10 others, 2007, A large genomic deletion in the PDHX gene caused by the retrotranspositional insertion of the full-length LINE-1 element, *Human Mutation* **28**, 137–142.

23. Chen, S. M., P. D. Stenson, D. N. Cooper, and C. Ferec, 2005, A systematic analysis of LINE-1 endonuclease-dependent retrotranspositional events causing human genetic disease, *Human Genet.* **117**, 411–427. There is also independent genetic evidence that many LINE-1 retroelements are under negative selection in *Homo sapiens*, implying that they have a substantial negative impact on human genetic fitness (Boissinot, S., J. Davis, A. Entezam, D. Petrov, and A. V. Furano, 2006, Fitness cost of LINE-1 (L1) activity in humans, *Proc. Natl. Acad. Sci, USA* **103**, 9590–9594).

24. Burwinkle, B. and M. W. Kilimann, 1998, Unequal homologous recombination between LINE-1 elements as a mutational mechanism in human genetic disease, *J. Mol. Biol.* **277**, 513–517; Segal, Y., B. Peissel, A. Renieri, M. de marchi, A. Ballabio, Y. Pei, and J. Zhou, 1999, LINE-1 elements at the sites of molecular rearrangements in Alport syndrome-diffuse leiomyomatosis, *Amer. J. Human Genet.* **64**, 62–69; Temtamy, S. A. and 12 others, 2008, Long interspersed nuclear element-1 (LINE1)-mediated deletion of EVC, EVC2, C40rf6, and STK32B in Ellis-van Creveld syndrome with borderline intelligence, *Human Mutat.* **29**, 931–938.

25. Hedges, D. J. and P. L. Deininger, 2007, Repetitive elements and human disorders, *Encyclopedia of Life Sciences*, John Wiley, New York.

26. This might be especially true for insertions of many of the larger mobile elements, such as full-length *L1*s, where the disruptive effects on genomes might often be greatest.

27. Several such examples are described in footnote 23. Another type of benefit to the host is illustrated by the second-order effects of

homologous recombination between mobile elements, such as *L1* sequences. In mammals, homologous recombination is often involved in a cell's repair of double-strand breaks (DSBs) in DNA. To the extent that *L1* sequences promote homologous recombination, they also may facilitate DSB repair that can be important for genetic stability and cell survival (Han, K., J. Lee, T. J. Meyer, P. Remedios, L. Goodwin, and M. A. Batzer, 2008, LI recombination-mediated deletions generate human genomic variation, *Proc. Natl. Acad. Sci. USA* **105**, 19366–19371.

28. An exaptation is a biological adaptation in which the current biological function of a trait in question originally evolved for reasons unrelated to that trait's current role. Stephen J. Gould and Elizabeth Vrba coined the word in 1982 (Exaptation—a missing term in the science of form, *Paleobiology* **8**, 4–15).

29. Dawkins, R., *The Selfish Gene* (Oxford University Press, Oxford, UK, 1976). The selfish gene concept itself was originated by the late William Hamilton, who formally showed how injurious genes can evolve in sexual species by natural selection acting at the level of the gene (for a review see Hamilton, W. D., 2001, *Narrow Roads of Gene Land, Vol. 2, The Evolution of Sex,* Oxford University Press, Oxford, UK).

30. *Genes in Conflict* (Harvard University Press, Cambridge, MA, 2006).

31. Some of the passages in this section are little modified from those in the following review by the author: Natural history of the sexual genome: proactive drivers and passive drifters (2008, *Perspectives in Biology and Medicine* **51**, 484-489). For another popularized account of metaphorical constructs for the human genome, see the following by Avise: Evolving genomic metaphors: a new look at the language of DNA, *Science* **294**, 86–87, 2001.

32. Lynch, M., *The Origins of Genome Architecture,* Sinauer, Sunderland, MA, 2007.

CHAPTER 5. INTELLIGENT OR NON-INTELLIGENT DESIGN?

1. Behe recently wrote another trade book (*The Edge of Darwinism: The Search for the Limits of Darwinism*; 2007, Free Press, New York) in which he admits that natural selection can produce evolutionary change within species but that "limits of Darwinism" are reached somewhere between the taxonomic levels of species and orders (at which point a highly intelligent directive force must be invoked for some if not all evolutionary transitions). Behe accepts common descent for all life forms on Earth but argues that nonrandom mutations directed by a purposeful agent are required to generate new species or higher taxa.

2. See Avise, J. C., 2008, The natural history of the sexual genome: proactive drivers and passive drifters, *Perspect. Biol. Medicine* **51**, 484–489.

3. According to the classification scheme by Austin Burt and Robert Trivers (in *Genes in Conflict*), any piece of DNA in a sexual species could in principle benefit itself (increase the proportion of gametes, and thereby of zygotes of the next generation, in which it is represented) by adopting any of at least three proactively selfish tactics: interference, in which the selfish allele disrupts or sabotages the transmission of an alternate allele; overreplication, in which the selfish allele biases its intergenerational transmission by getting itself replicated more often than other alleles in the same host; and gonotaxis, by which the selfish allele moves preferentially toward the germ line. All three of these modes of genetic selfishness have been documented in the genomes of various sexual species.

4. Under one common evolutionary interpretation, the simplicity and functional efficiency of prokaryotic genomes (compared to eukaryotic genomes) are a result of at least two major factors. First, as mentioned in the text, many prokaryotic lineages are primarily asexual for much of their existence, whereas most eukaryotic species are obligately sexual. The evolutionary proliferation of mobile elements is normally possible only in sexual species because only in such species are the genetic fates of various pieces of DNA effectively decoupled such that selfish genes can profit by self-proliferation (e.g., Hickey, 1982, Selfish DNA: a sexually-transmitted nuclear parasite, *Genetics* **101**, 519–531). Second, most bacterial populations are huge, so their genomic features are typically far more visible to natural selection than is true in most eukaryotes (where smaller population sizes allow greater opportunities for random genetic drift to override the effects of natural selection). Finally, another relevant point is that some multicellular organisms, including about 100 vertebrate species, are clonal (e.g., parthenogenetic) rather than sexual (Avise, J. C., 2008, *Clonality: The Genetics, Ecology, and Evolution of Sexual Abstinence in Vertebrate Animals*, Oxford University Press, New York). Such creatures undoubtedly do contain legions of mobile elements but presumably only for reasons of history legacy (nearly all such clonal taxa are known to be recent evolutionary derivatives from sexual ancestors).

5. The metaphors described in the text are developed more fully in Avise, J. C., 2001, Evolving genomic metaphors: a new look at the language of DNA, *Science* **294**, 86–87.

6. A nontechnical summary of the biochemistry of blood-clotting can be found on pp. 152–161 in Miller, K. R., 1999, *Finding Darwin's God*, HarperCollins, New York.

7. Liu, R. and H. Ochman, 2007, Stepwise formation of the bacterial flagellar system, *Proc. Natl. Acad. Sci. USA* **104**, 7116–7121.

8. Miller, K. R., 2004, The flagellum unspun: the collapse of "irreducible complexity," pp. 81–97 in *Debating Design: From Darwin to DNA*, W. Dembski and M. Ruse (eds.), Cambridge University Press, Cambridge, UK); Musgrave, I., 2004, Evolution of the bacterial flagellum, pp. 48–84 in *Why Intelligent Design Fails*, M. Young and T. Edis (eds.), Rutgers University Press, New Brunswick, NJ); Pallen, M. J. and N. J. Matzke, 2006, From *The Origin of Species* to the origin of the bacterial flagella, *Nature Rev. Microbiol.* **4**, 784–790).

9. For readers concerned that evolution has no analogue to the havens of safety proposed in the highway metaphor, the median strips are not a necessary part of the scenario. Imagine instead the running groundhogs spew out progeny continuously as they travel across the lanes of traffic. Mortality would be high, but at least some progeny and their descendants could plausibly reach the far side of the road.

10. Dobzhansky, T., 1967, *The Biology of Ultimate Concern*, New American Library, New York (as also quoted in Ayala, F. J., 2007, *Darwin's Gift to Science and Religion*, Joseph Henry Press, Washington, DC).

11. Proponents of Intelligent Design often argue that we are in no position to judge a Creator God's products as error-ridden because we cannot know the method or the real purposes of his works. This stance is self-contradictory, of course, if at the same time we are allowed to interpret purported phenomena such as "irreducible complexity" as evidence for a purposeful intelligent design.

12. Admittedly, scientists will never be able to prove the null hypothesis that a given piece of DNA serves absolutely no utilitarian role in a cell because a skeptic can always contend that some positive function remains to be discovered. In any event, the hypothesis that many features of the human genome are nonadaptive for their bearers has been valuable because it has provided a strong stimulus for scientific research in molecular genetics.

13. For readers concerned about the loss of intron-enabled alternative splicing as a source of protein diversity during ontogeny, consider the following. If introns, mobile elements, and other such pieces of "superfluous" DNA were somehow jettisoned from the human genome, at least 100-fold more protein-coding loci could be housed without an increase in overall genome size. This would be far more than enough space to accommodate genes for all of a cell's required proteins, including those now generated by alternative splicing.

14. For readers concerned that an intelligent agent had some good reason for imbuing each human cell with two genomes (mitochondrial

plus nuclear) rather than one, it could be pointed out that under intelligent design, presumably these two collaborating genomes should at least be able to cross-communicate effectively. In fact, however, the mitochondrial genetic code differs in details from the otherwise universal genetic code employed by the nuclear genome. For example, AGA and AGG are stop codons in vertebrate mtDNA but not in nuclear DNA. These differences in code are one reason that messenger RNAs from mtDNA cannot be translated effectively by the ribosomes employed by nuclear loci, and vice versa.

15. Under an evolutionary interpretation (the endosymbiotic theory), the human mitochondrial genome is ultimately of bacterial descent. The presence of such a streamlined genome within human cells demonstrates that there is no insuperable engineering hurdle if an intelligent designer had decided from the outset to likewise imbue the human nuclear genome with these molecular refinements.

16. Bloom, J. D. and F. H. Arnold, 2009, In the light of directed evolution: pathways of adaptive protein evolution, pp. 9995–10000 in *In the Light of Evolution, III, Two Centuries of Darwin,* J. C. Avise, and F. J. Ayala (eds.), Proc. Natl. Acad. Sci. USA **106** (supplement 1), 9933–10066.

17. Aiuti, A. and colleagues, 2009, Gene therapy for immunodeficiency due to adenosine deaminase deficiency, *New England J. Medicine* 360: 447–458.

18. For the general reader, I can recommend the following accessible treatment (and references therein): Avise, J. C., 2004, *The Hope, Hype, and Reality of Genetic Engineering: Case Histories from Agriculture, Industry, Medicine, and the Environment,* Oxford University Press, Oxford, UK.

19. Some Christians might suggest that genetic engineering by humans is itself consistent with God's design and intentions, when we use our intelligence—that which makes us in the image of God—to overcome some of the terrible natural effects of biological evolution. In other words, given that horrendous biological conditions do exist in our species, to overcome them could be interpreted as God's work.

20. 2008, University of California Press.

21. Ayala, F. J. 2008. From Paley to Darwin: design to natural selection, pp. 50–75 in *Back to Darwin: A Richer Account of Evolution,* J. B. Cobb, Jr. (ed.), William B. Eerdman, Grand Rapids, MI.

22. More than 200 years ago, a similar sentiment was expressed by the philosopher David Hume when he wrote, "Is he [God] willing to prevent evil, but not able? Then he is impotent. Is he able, but not willing? Then he is malevolent. Is he both able and willing? Whence the evil?" This statement actually has much deeper historical roots, tracing back to the Greek philosophic Epicurus three centuries B.C. (as

explained on p. 215 of *Darwin's Gift to Science and Religion*; see footnote 23 below).

23. *Darwin's Gift to Science and Religion* (2007, Joseph Henry Press, Washington, DC) is the title of an eloquent book by Francisco Ayala that develops this thesis at length. The idea is developed also by Jack Haught in "Darwin's gift to theology" (1998, pp. 393–418 in *Evolutionary and Molecular Biology: Scientific Perspectives on Divine Action*, R. J. Russell, W. R. Stoeger, and F. J. Ayala [eds.], Vatican City State).

24. All of the excerpts are requoted from *Science, Evolution, and Creationism*, 2008, National Academy of Sciences and the Institute of Medicine, National Academies Press, Washington, DC.

25. It might be noted that the current Pope Benedict seems more measured or cautious on this point and might not necessarily agree entirely with Pope John Paul II.

Epilogue

1. *Chambers English Dictionary*, Cambridge.

Additional Readings

I highly recommend the following books (among many others currently available) to readers interested in further thought about the evolution/religion interface from the perspectives of the history or philosophy of science.

Avise, J. C. 1998. *The Genetic Gods: Evolution and Belief in Human Affairs*. Harvard University Press, Cambridge, MA. The author develops an extended metaphor in which genes (the genetic gods) and their protein angels are interpreted to exercise, over many human affairs, influences that philosophers and theologians traditionally had reserved for supernatural deities. Avise contends that modern genetic knowledge can thereby inform our attempts to answer typically religious questions—about origins, fate, and meaning.

Ayala, F. J. 2007. *Darwin's Gift to Science and Religion*. Joseph Henry Press, Washington, DC An eloquent and insightful book for a general audience, developing the thesis that the Darwinian revolution completed the Copernican revolution by extending to biology the concept of nature as a lawful system of matter in motion that science can comprehend without recourse to supernatural causation. Ayala discusses reasons that Darwin's insights should be interpreted as a magnificent gift not only to science but to religion and philosophy as well.

Cobb, J. B., Jr. (ed.). 2008. *Back to Darwin: A Richer Account of Evolution.* William B. Eerdmans, Grand Rapids, MI. An edited volume with erudite chapters from leading scientists, philosophers, and theologians who generally attempt to reconcile or at least accommodate mainstream evolutionary science with Christian belief.

Dembski, W. A. and M. Ruse (eds.). 2004. *Debating Design: From Darwin to DNA.* Cambridge University Press, Cambridge, UK. An edited volume with contributions from a stellar lineup of evolutionary biologists, historians, and philosophers, and also including leading proponents of intelligent design. This compilation purports to give a comprehensive and "even-handed" overview of current perspectives on the origins of biological design.

Dennett, D. C. 1995. *Darwin's Dangerous Idea: Evolution and the Meanings of Life.* Simon & Schuster, New York. A masterful work by an eloquent spokesperson for the power of natural selection and for how Darwin's grand ideas have transformed our sense of place in the universe. Dennett contends that far from destroying meaning in life, Darwinism puts issues of ultimate concern on a foundation that not only is more solid scientifically but also in many ways more inspirational.

Forrest, B. and P. R. Gross. 2004. *Creationism's Trojan Horse: The Wedge of Intelligent Design.* Oxford University Press, Oxford, UK. A critique on the agenda of the Intelligent Design movement, this book is as meticulously researched as it is revealing and devastating.

Kurtz, P. (ed.). 2003. *Science and Religion: Are They Compatible?* Prometheus Books, Amherst, NY. A fascinating edited collection of chapters by leading scientists expressing their personal worldviews on the relationship and common ground (if any) between science and religion.

Miller, K. R. 1999. *Finding Darwin's God: A Scientist's Search for Common Ground between God and Evolution.* HarperCollins, New York. In developing an unorthodox resolution to the evolutionism-creationism debate, Miller demolishes various claims of evolution's critics and then goes on to address the philosophical ramifications. Does evolution invalidate spiritual worldviews? Does it demand agnosticism, or perhaps even atheism? Miller's answers to these questions are "no" and "no."

Ruse, M. 2001. *Can a Darwinian Be a Christian?* Cambridge University Press, New York. Michael Ruse is a superb historian and philosopher of science. His considered answer to the question posed in the book's title is a resounding, yet qualified, "yes." Other relevant books authored by Ruse include *Mystery of Mysteries—Is Evolution a Social Construction?* (1999), and *Darwin and Design—Does Evolution Have a Purpose?* (2003), both published by Harvard University Press, Cambridge, MA.

Sarkar, S. 2007. *Doubting Darwin? Creationist Designs on Evolution.* Blackwell, Malden, MA. Sarkar argues for a broadly naturalistic framework for understanding biology, and in the process explains why most biologists find objectionable the ID brand of creationism.

Scott, E. C. 2004. *Evolution vs. Creationism.* Greenwood Press, Westport, CT. An encyclopedic compendium of well-researched information on the history, conceptual and empirical pillars, and ramifications of the evolutionary sciences versus creationism in society.

Sober, E. 2008. *Evidence and Evolution: The Logic behind the Science,* Cambridge University Press, Cambridge, UK. This is a deep look at the epistemology and philosophy of the evolutionary sciences and also of the hypothesis of intelligent design. It includes some unorthodox but also compelling criticisms of the ID paradigm.

Thomson, K. 2005. *The Watch on the Heath: Science and Religion before Darwin.* HarperCollins, London. A superb history of natural theology and of natural theologians in the pre-Darwinian era. This treatment illuminates the intellectual struggles of leading historical figures who attempted to reconcile their belief systems with the science of their times.

Wilson, D. S. 2002. *Darwin's Cathedral: Evolution, Religion, and the Nature of Society.* University of Chicago Press, Chicago, IL. A thoughtful and provocative evolutionary argument developing the thesis that morality and religion are biologically and culturally evolved adaptations that have promoted the formation and perpetuation of human social groups. Wilson contends that religion has enabled people to accomplish by cohesive and collective action what could not otherwise have been achieved.

GLOSSARY

Adaptation. Any feature (e.g., morphological, physiological, behavioral) that makes an organism suited to survive and reproduce in a particular environment.

Aging. (See senescence.)

Allele. Any of the possible alternative forms of a gene. A diploid individual carries two alleles at each autosomal gene, and these can either be identical in state (in which case the individual is homozygous) or different in state (heterozygous). At each autosomal gene, a population of N diploid individuals harbors 2N alleles, many of which may differ somewhat in nucleotide sequence.

Alternative splicing. The ligation of exons of a particular gene to form a functional RNA that differs in information content from the normal messenger RNA.

Amino acid. One of the molecular subunits polymerized to form polypeptides.

Antagonistic pleiotropy. A form of pleiotropy in which a gene's influence on one phenotypic trait is beneficial to the organism but its influence on another trait is detrimental.

Asexual reproduction. A reproductive mode that does not involve the fusion of sex cells.

Autosome. A chromosome in the nucleus other than a sex chromosome; in diploid organisms, autosomes are present in homologous pairs. (See also sex chromosome).

Bacterium. A unicellular microorganism without a true cellular nucleus.

Biochemistry. The chemistry of life.

Blasphemy. Contempt of or indignity to God.

Bottleneck. (See population bottleneck.)

Cancer. A disease characterized by uncontrolled cellular proliferation.

Carbohydrate. An organic compound (including sugars and starches) composed of a chain or ring of carbon atoms to which hydrogen and oxygen atoms are attached in a ratio of approximately 2:1.

Cell. A small, membrane-bound unit of life capable of self-reproduction.

Chromatin. The complex of DNA and proteins of which eukaryotic chromosomes are composed.

Chromosome. A threadlike structure within a cell that carries genes.

Clade. A monophyletic assemblage; a group of species (or, sometimes, conspecific individuals) that share a closer common ancestry with one another than with any other such group.

Codominance. A genetic situation in which both alleles in a heterozygous diploid individual are expressed simultaneously in the phenotype.

Complementary DNA (cDNA). DNA produced from an RNA template by a reversal of transcription.

Conspecific. Belonging to the same species.

Creationism. A shorthand for the sentiments of the Creation Science movement.

Creation Science. A religious movement with the doctrine that a supernatural Deity engineered life directly—i.e., not through evolutionary processes—and recently.

Cytoplasm. The portion of a eukaryotic cell outside of the nucleus.

Deleterious. In reference to genetics, any genetic condition that harms an organism's health.

Deletion. The loss of a segment of genetic material from a chromosome.

Deoxyribonucleic acid (DNA). A double-stranded molecule each of whose nucleotide subunits is composed of a deoxyribose sugar, a phosphate group, and one of the nitrogenous bases adenine, guanine, cytosine, or thymine.

Diploid. A usual condition of a somatic cell in which two copies of each chromosome are present. (See also haploid.)

Dominant allele. An allele that, in a heterozygous diploid individual, is expressed fully in the phenotype. (See also recessive allele.)

Drift. (See genetic drift.)

Duplication. A genetic event usually stemming from an abnormal meiosis in which a gene or portion of a chromosome gives rise to a second copy.

Duplicon. Any of the duplicated copies of a protein-coding gene.

Ecosystem. A community of organisms and the physical environment with which it interacts.

Embryo. An organism in the early stages of development (in humans, usually up to the beginning of the third month of pregnancy).

Endosymbiosis. A form of symbiosis in which one organism lives within the body of another.

Enzyme. A protein that catalyzes a specific chemical reaction.

Epigenetics. Changes in gene expression that are inherited but not caused by changes in DNA sequence per se. In a broader sense, epigenetics can also refer to the entire suite of mechanisms, developmental pathways, and social and other environmental influences by which genomes give rise to organismal-level features.

Erythrocyte. A red blood cell.

Eukaryote. Any organism in which chromosomes are housed in a membrane-bound nucleus.

Evolution. Change through time in the genetic composition of a population.

Exaptation. A biological adaptation in which the current biological function of a trait in question originally evolved for reasons unrelated to that trait's current role.

Exon. A gene segment that codes for a polypeptide. (See also intron.)

Fetus. An organism in intermediate stages of development in the uterus (in humans, beginning at about the third month of pregnancy).

Fitness (genetic). The contribution of a genotype to the next generation relative to the contributions of other genotypes in the population.

Flagellum. The "tail" of a bacterium.

Fossil. Any remain or trace of life no longer alive.

Frameshift. Any mutation in protein-coding DNA that results in a shift in the reading frame by which successive triplets of nucleotides are transcribed and translated into a polypeptide.

Fundamentalism. Belief in the literal truth of the Bible (or other such religious text or source).

Gamete. A mature reproductive sex cell (egg or sperm).

Gametogenesis. The specialized series of cellular divisions that leads to the production of gametes. (See also meiosis, oogenesis, spermatogenesis.)

Gene. The basic unit of heredity; usually taken to imply a sequence of nucleotides specifying production of a polypeptide or other functional product such as ribosomal RNA, but also can be applied to stretches of DNA with unknown or unspecified function.

Genealogy. A record of descent from ancestors through a pedigree.

Gene pool. The sum total of all hereditary material in a population or species.

Gene therapy. The human-mediated insertion of a functional gene into an individual's cells with the intent of correcting a hereditary disorder.

Genetic code. The consecutive nucleotide triplets of DNA and RNA that specify particular amino acids for protein synthesis.

Genetic drift. Change in allele frequency in a finite population by chance sampling of gametes from generation to generation.

Genetic engineering. Any experimental or industrial method employed to alter the genomes of living cells.

Genetic load. The collective burden of genetic defects in a population.

Genetics. The study of heredity and hereditary molecules.

Genome. The complete genetic constitution of an organism; also can refer to a particular composite piece of DNA, such as the mitochondrial genome.

Genotype. The genetic constitution of an individual organism with reference to a single gene or set of genes. (See also phenotype.)

Germ cell. A sex cell or gamete. (See also somatic cell.)

Germ line. Pertaining to the cellular lineage from which germ cells are derived.

Haploid. A usual condition of a gametic cell in which only one copy of each chromosome is present. (See also diploid.)

Heredity. Genetic inheritance; the phenomenon of familial transmission of genetic material from generation to generation.

Heritability. The fraction of variation of a trait within a population due to heredity as opposed to environmental influences.

Heterogametic sex. The gender that produces gametes containing unlike sex chromosomes (in humans, the male). (See also homogametic sex.)

Heteroplasmy. The co-occurrence of two or more cytoplasmic genotypes (such as those for mtDNA) within a cell or individual.

Heterosis. The condition in which heterozygotes have higher genetic fitness than homozygotes.

Heterozygote. A diploid organism possessing two different alleles at a specified gene. (See also homozygote.)

Homogametic sex. The gender that produces gametes containing alike sex chromosomes (in humans, the female). (See also heterogametic sex.)

Homology. Similarity of structure due to inheritance from a shared ancestor; can refer to any structural features ranging from DNA sequences to morphological traits.

Homozygote. A diploid organism possessing two identical alleles at a specified gene. (See also heterozygote.)

Hormone. An organic compound produced in one region of an organism and transported to target cells in other parts of the body where its effects on phenotype are exerted.

Imprinting, genetic. Mechanisms (including methlyation) by which the expression of an allele depends on whether the allele came from the individual's male or female parent.

Independent assortment (Mendel's Law of). The random distribution to gametes of alleles from genes on different chromosomes, or from genes far enough apart on a given chromosome.

Insertion. The entry of a new piece of DNA into a chromosome.

Intelligent Design (ID). A recent incarnation of the Creation Science movement in which a supreme intelligence is invoked to account for biological complexity and perfection.

Intron. A noncoding portion of a gene. Most genes in eukaryotes consist of alternating intron and exon DNA sequences.

Inversion. A genetic condition in which a chromosomal segment has been rotated 180° from its original linear orientation.

Irreducible complexity. The notion (promoted by Michael Behe) that a complex biological trait exhibits a total loss of function if any of its parts is removed, the implication being that such a trait therefore must have been created in toto by an intelligent designer.

Jumping gene. (See transposable element.)

Junk DNA. DNA that contributes nothing beneficial to the organism in which it is housed.

Kinase. An enzyme that phosphorylates (adds phosphate groups to) and thereby alters the activity of a substate molecule.

Lipid. Biochemical substances (including fats, oils, and waxes) that are barely soluble in water but variably soluble in organic solvents such as alcohol.

Linkage, genetic. The co-occurrence of particular loci on the same chromosome, thus often implying (if the loci are close enough) some restriction on recombination between them.

Locus (pl. loci). A gene, or a specified stretch of DNA.

Meiosis. The cellular process whereby a diploid cell divides to form haploid gametes.

Messenger RNA. A form of ribonucleic acid transcribed from structural genes, the exon-derived portions of which subsequently will be translated into a polypeptide.

Metabolic pathway. A series of stepwise biochemical changes in the conversion of a precursor substance to an end product, each step typically catalyzed by a specific enzyme.

Metabolism. The sum of all physical and chemical processes by which living matter is produced and maintained, and by which cellular energy is made available.

Metaphysical. Pertaining to metaphysics, i.e., to phenomenological explanations that tend to be based on the supernatural, magical, occult, or transcendental.

Microbe. An organism too small to be seen with the unaided eye.

Micro-RNA (miRNA). A short stretch of RNA that can bind to complementary sequences in a messenger RNA molecule and thereby inhibit the translation or induce the degradation of a specific genetic message.

Microsatellite locus. A stretch of DNA containing short repeated sequences each typically about two to six base-pairs in length.

Mitochondrion. An organelle in the cell cytoplasm that contains its own DNA (mtDNA) and that is the site of some of the metabolic pathways involved in cellular energy production.

Mitosis. A process of cell division that produces daughter cells with the same chromosomal constitution as the parent cell. (See also meiosis.)

Mobile element. (See transposable element.)

Mutagenic. Mutation-inducing.

Mutation. A spontaneous change in the genetic constitution of an organism.

Natural selection. The differential survival and reproduction of individuals with different genotypes.

Natural Theology. The idea that the beauties and workings of nature provide final and definitive proof of God's majesty.

Neutral mutation. A mutation that neither enhances nor diminishes genetic fitness.

Nonsense mutation. A mutation that generates a stop codon where it does not belong.

Nonsynonymous mutation. A nucleotide substitution in DNA that results, by virtue of the genetic code, in a different amino acid being incorporated into the translated polypeptide.

Nucleic acid. (See deoxyribonucleic acid or ribonucleic acid.)

Nucleotide. A unit of DNA or RNA consisting of a nitrogenous base, a pentose sugar, and a phosphate group.

Nucleus. The portion of a eukaryotic cell bounded by a nuclear membrane and containing chromosomes.

Oocyte. An egg cell.

Oogenesis. The specialized series of cellular divisions that leads to the production of haploid oocytes.

Organ. A recognizable body feature (such as the heart or kidney) composed of several different tissues grouped together into a functional and structural unit.

Parasite. An organism that lives on or in an organism of a different species and derives nutrients from its host.

Phenotype. The observable properties of an organism at any level, ranging from molecular and physiological to gross morphological.

Philosophy. The pursuit of wisdom and knowledge about the nature of all things.

Phylogeny. Evolutionary relationships (historical descent) of a group of organisms or species.

Pleiotropy. A genetic phenomenon wherein a single gene or genetic alteration influences multiple phenotypic features.

Point mutation. A mutation at a particular nucleotide position or location in a gene.

Polygenic trait. A phenotypic trait affected by multiple genes.

Polymerase. An enzyme that catalyzes the formation of nucleic acid molecules.

Polymorphism. With respect to genetics, the presence of two or more genotypes in a population.

Polypeptide. A polymer composed of amino acids chemically linked together.

Polyphyletic. Having multiple independent evolutionary origins.

Population bottleneck. A severe but often temporary reduction in the size of a population.

Population genetics. The study of evolutionary forces that can change the genetic composition of populations.

Preadaptation. (See exaptation.)

Prokaryote. Any microorganism that lacks a chromosome-containing, membrane-bound nucleus.

Protein. A biological macromolecule composed of one or more polypeptide chains.

Protozoan. A unicellular animal.

Pseudogene. A gene bearing close structural resemblance to a known functional gene at another chromosomal site but that itself is nonfunctional due to genetic alterations such as additions, deletions, or nucleotide substitutions.

Reading frame. The phase in which successive triplets of a nucleic acid sequence are translated.

Recessive allele. An allele that, in a heterozygous diploid individual, is masked in phenotypic expression by a dominant allele at the same locus. (See also dominant allele.)

Recombinant DNA. A new DNA molecule that has arisen from genetic recombination (often mediated by humans).

Recombination (genetic). The formation of new combinations of genes through such natural processes as meiosis and fertilization, or in the laboratory through recombinant DNA technologies.

Regulatory gene. A gene that exerts operational control over the expression of other genes.

Religion. Any system of belief and worship, typically coupled with emotion and morality, of a higher controlling power or powers.

Retrotransposable element. A mobile element that moves about the genome via an intermediate RNA molecule which then is reverse-transcribed into DNA.

Retrotransposon. (See retrotransposable element.)

Retrovirus. An RNA virus that utilizes reverse transcription during its life cycle to integrate into the DNA of a host cell.

Ribonucleic acid (RNA). A single-stranded polynucleotide molecule each of whose nucleotide subunits is composed of a ribose sugar, a phosphate group, and one of the nitrogenous bases adenine, guanine, cytosine, or uracil.

Ribosomal RNA. A form of ribonucleic acid that together with ribosomal proteins composes a ribosome.

Ribosome. An organelle in the cell cytoplasm composed of RNA and protein and that is the site of protein translation.

Science. Objective knowledge, or the means by which such knowledge is attained, through careful observation, experiment, and logical inference.

Segregation (Mendel's Law of). The distribution to gametes of the two alleles in a diploid individual; each gamete receives, at random, one or the other of the two alleles at each gene.

Selfish DNA. DNA that displays self-perpetuating modes of behavior without apparent benefit to the organism.

Senescence. A persistent decline in the age-specific survival probability or reproductive output of an individual due to internal physiological deterioration.

Sex chromosome. A chromosome in the cell nucleus involved in distinguishing the two genders. In humans, the "X" and "Y" are sex chromosomes. (See also autosome.)

Sexual reproduction. Reproduction involving the production and subsequent fusion of haploid gametes.

Sexual selection. The differential ability of individuals of the two genders to acquire mates. The topic can be subdivided into two components: intrasexual selection, which refers to competition among members of the same sex over access to mates; and intersexual or epigamic selection, which refers to mating choices made between males and females.

Signal transduction. Mechanistic pathways by which chemical or other environmental stimuli evoke cellular responses.

Sister taxa. One another's closest phylogenetic (evolutionary) relatives.

Somatic cell. Any cell in a eukaryotic organism other than those destined to become germ cells. (See also germ cell.)

Spermatogenesis. The specialized series of cellular divisions that leads to the production of haploid sperm cells.

Spliceosome. A large ribonucleoprotein that biochemically removes all intron-derived segments from each pre-mRNA and then splices each gene's exons end-to-end to generate a mature mRNA.

Stochastic. Chancy; governed largely or exclusively by random events.

Stop codon. A codon that acts as a signal for the termination of the translation process.

Substitution. A mutational change from one type of nucleotide to another at some position along a nucleic acid.

Symbiosis. A close interaction or association (usually implied to be mutually beneficial) between organisms belonging to different species.

Syngamy. The union of two gametes to produce a zygote; fertilization.

Synonymous mutation. A nucleotide substitution in DNA that does not result (by virtue of redundancy in the genetic code) in a different amino acid becoming incorporated into the translated polypeptide.

Systematics. The study of evolution and classification of organisms into hierarchical series of groups.

Taxonomy. The theory and practice of naming and classifying organisms.

Thallasemia. A category of hemoglobinopathies (disorders related to globin molecules) resulting from metabolic disruptions in the rate of production (rather than the biochemical structures per se) of hemoglobin molecules.

Theodicy. Vindication of the justice of God in establishing a world in which evil exists.

Theology. The study of God, religion, and revelation.

Tissue. A biological feature composed of similar cells organized into a functional and structural unit.

Transcription. The cellular process by which an RNA molecule is formed from a DNA template.

Transfer RNA. A form of ribonucleic acid that picks up amino acids from the cell cytoplasm and moves them into position for the translation process.

Translation. The process by which the genetic information in messenger RNA is employed by a cell to direct the construction of polypeptides.

Translocation. An interchange of chromosomal segments between nonhomologous chromosomes, or between distant regions of homologous chromosomes.

Transposable element. Any of a class of DNA sequences that can move from one chromosomal site to another, often replicatively.

Transposition. The process by which a replica of a transposable element is inserted into another chromosomal site.

Virus. A tiny, obligate intracellular parasite incapable of autonomous replication but which instead utilizes the host cell's replicative machinery.

X-chromosome. The sex chromosome normally present as two copies in female mammals (the homogametic sex), but as only one copy in males (the heterogametic sex).

Y-chromosome. In mammals, the sex chromosome normally present in males only.

Zygote. The diploid cell arising from the union of male and female haploid gametes.

INDEX

Page numbers followed by *f* denote figures; those followed by *t* denote tables.